PHYSICS 11–13

Longman Science 11–13 Series
General Editor: John L. Lewis

Also in this series: **Chemistry 11–13**

PHYSICS 11–13

John L. Lewis,
Senior Science Master, Malvern College

LONGMAN

Longman Group Limited
London
*Associated companies, branches and representatives
throughout the world*

©Longman Group Ltd 1977

First published 1977
Second impression 1978

ISBN 0 582 20877 7

Printed in Great Britain by Butler & Tanner Ltd, Frome & London

Preface

This book is a course in Physics for pupils aged 11–13. It uses a modern experimental approach to impart an *understanding* of the principles of physics. Considerable care has been taken to ensure that both the approach and the writing are suitable for pupils with a range of abilities.

It is not easy to write such a book if the text is not to ruin the spirit of enquiry for it would be unfortunate if it provided all the answers to the experiments which it is intended the pupils would do for themselves. It is hoped that the book has avoided this without leaving so much open-ended that the pupils have no idea where they are going.

Pupils certainly learn much through careful questions and, for that reason, there are a very large number included in the text and, at various stages, there are revision tests which teachers should find useful.

The physical ideas and concepts developed in this course are fundamental whatever course the pupils will do next. There is a temptation to include a great deal of background material of general interest, which would help to show the relevance of the work to the outside world. Although references are included, the temptation has been resisted to some extent as such material is contained in the series of Longman Physics Topics, which continue to provide most suitable background material and which it is hoped will be used to supplement this text book.

The author owes much to the wealth of ideas from many sources which were fed into the Nuffield course and which have clearly affected his own ideas. Above all, he owes a debt of gratitude to his friends and colleagues Geoffrey Foxcroft, Alan Duff, Alec Porch, Jim Brown and the Reverend Michael Phillips, who have helped him both on the Common Entrance Science panel and over the writing of this book.

Acknowledgements

The author and publisher are grateful to the following for permission to reproduce photographs: Bently W. A. and Humphreys W. J., *Snow Crystals* (McGraw Hill), page 22 (top right); Paul Brierley, page 154; British Rail, page 150; British Steel Corporation, page 10; Bruce Coleman Ltd, page 225; Camera Press Ltd, page 100 (bottom); Central Electricity Generating Board, page 158 (top); Dunlop Ltd, page 70; English Tourist Board, page 158 (bottom); French Government Tourist Office, page 159 (left); Institute of Geological Sciences, London, page 15; Keystone Press Agency, page 148 (centre left & bottom right); Martin, *Thirteen Steps to the atom* (Harrap), page 16 (right) and 21; Popperfoto, page 100 (top) and 148 (bottom left); RHM Research Ltd, page 14 (bottom); Science Museum, London, page 155 and 156 (bottom right); Swiss National Tourist Office, page 159 (right); Tate & Lyle Ltd, page 14 (top left); United Kingdom Atomic Energy Authority, page 156 (top left and right) and 217; White Electrical Instrument Co Ltd, page 197 (centre left and right);

The author is also grateful to the Common Entrance Committee for permission to reproduce a large number of questions which have been used in Common Entrance physics papers.

Contents

electrical leads

bicycle frame

glass window

bricks and mortar

chairs and table of wood

car bumper

iron girder bridge

saucepans

Cotswold stone house

aeroplane

mattress and pillow

milk bottle and glasses

hammer and nails

curtains

mercury thermometer

pylons and cable

porcelain

car tyres

paper

ship's cable

lubricating oil

blanket, rug and sheets

polythene bowls

balloon

Chapter 1 **Materials**

The variety of materials

In our homes, in our schools, in the streets and in the countryside we are surrounded by all kinds of things. There is also an immense variety in the materials of which these things are made. In your home there are tables and chairs made of wood, saucepans of aluminium, windows of glass, curtains of cotton, bowls of polythene, bicycle frames of iron, electrical wires of copper.

In your laboratory you will have seen a variety of different kinds of materials. Some of the substances were *solid*: iron, copper, aluminium, rubber, brick, lead, wood, polythene. Some of the solids consisted of little bits, but the bits were none the less solid: sugar, salt, iron filings, sand. These substances have very varied properties. Some are hard, some soft; some bend easily, some break as soon as you bend them; some stretch; some can be dented by your finger, some spring back to their original shape when your finger is removed; they may differ in smell, in colour and in their feel or texture.

When we wish to make something we choose carefully the substance with the most suitable properties. Why do you think we make furniture from wood? Why do we not choose wood for a saucepan? Why use aluminium instead? Why is the handle made of a different substance from the rest of the saucepan? Why are tyres made of rubber? Why is rubber unsuitable for a saw?

Some of the substances you have seen at home or in your laboratory are *liquid*: water, paraffin, mercury, treacle, vinegar. They too have varied properties: some for example flow very easily, others are very sticky, some smell, some are coloured.

Yet other substances are *gases*, like the air around us. Perhaps you have seen balloons filled with different kinds of gases: a balloon of hydrogen will rise upwards, a balloon of carbon dioxide will fall.

Solid, liquid and gas

How can you tell whether a substance is a solid or a liquid? One sensible answer would be to see if it will pour. Oil will pour, but a block of iron does not, and we therefore call oil a liquid, but iron a solid. Or will iron pour? The picture on the left shows molten iron being poured.

Why does it pour in this case? Because it has been heated, and this leads us to another important property of matter. Many solids when heated will become hot enough for them to melt and turn into a liquid. Heat the liquid and it will eventually boil and turn into gas. The most familiar example of this is ice: heat it and it becomes water, a liquid; heat it further and the water turns to steam, a gas. Normally iron is a solid, but heated sufficiently it turns to liquid and any iron in a star like the Sun would be a gas.

Sometimes it is a little difficult to decide whether a substance is solid or liquid. A piece of pitch may appear to be solid, but if a tin of pitch is left on its side, it will eventually flow out. What about glass? If you examine old glass in a church you may find that the glass is thicker at the bottom than at the top. What could be the explanation of this?

Molten iron being poured

Questions for class discussion

Why are saucepans made of aluminium? Why are they not made of polythene or wood?

Why are electrical connections usually made with copper wire? Why do we not use silver? Why is the copper wire often surrounded with rubber or plastic?

Could the blade of a spade be made of lead or aluminium instead of steel?

Why are car bumpers made of steel? Why are they usually chromium plated?

Why are tyres made of rubber?

What material is used for making aeroplanes?

Why are ships made of iron? Since a piece of iron sinks when it is put on water, how is it possible for a ship made of iron to float?

What kind of material is useful for clothing on a cold day?

Would gold be a good substance for a garden fork? Would there be any advantages in an iron fork with a thin layer of gold over the iron (gold plated)? What are the disadvantages?

Why are most houses in the Cotswolds made of stone?

Why is concrete used in building bridges, but not for motor car bodies?

A game to play

Put on a tray a block of each of the following: iron, brass, lead, glass, aluminium, softwood, hardwood, foamed polystyrene, perspex, paraffin wax. Feel each of them. Weigh them in your hand. Smell them. Find how hard they are by trying to dig your finger nail into them. When you think you know them, get someone to blindfold you and to move the blocks around on the tray. Then see if you can decide which block is which. You will be doing well if you get them all right.

There is a more difficult version of this game. Start with ten blocks as before and get to know them in the same way. But when you are blindfolded try to identify ten other objects, made of the same materials, but different in shape. Some might be tubes, some rods, some blocks, some wires and so on. This is much harder and really tests how well you got to know the original blocks.

The above game is typical of what a scientist has to do. To decide what each article was made of you had to find clues — how hard it was, how strong it was and so on. A scientist is always looking for evidence (clues) and on the basis of that evidence forms an opinion. The following game is another which requires looking for clues. It is also a good one to try out on parents. See how good they are at looking for evidence.

Another game

Get a collection of old cocoa or coffee tins, preferably all the same shape and size. Put a different object inside each and then seal the tin with tape. Put a number on the outside of each tin. Then challenge someone to try to decide what is inside each tin by looking for clues.

The tins might contain some of the following: a rectangular block of aluminium, a rectangular block of lead, a golf ball, two golf balls, a cylinder of steel, iron nails, a little lead shot, a lot of lead shot, some water, some thick oil, a lemon, some sand, a flat metal disc, a rubber bung and so on.

Homework assignments

1. Read the Longman Physics Topics book *Materials*, pages 11–21, and prepare a short talk to the class about 'Stony Materials used in building'. Illustrate your talk with pictures or cuttings from newspapers, magazines or other sources.

2. Read *Materials*, pages 22–23, and make a list of metals. Write down the properties of each, and then list some of the uses to which each of them are put.

3. Read *Materials*, pages 34–41, and make a collection of as many different kinds of wood as you can. Large pieces are not necessary, but the collection might include balsa wood, some examples of hardwoods, some softwoods, some plywood, perhaps some chipboard or veneered wood. Prepare your collection for showing to the rest of the class.

4. Read the appropriate pages in *Materials* and prepare a short talk on one of the following: glass (pages 42–50), fibres: natural and man-made (pages 52–54), rubber (pages 55–57), plastics (page 58), ceramics (pages 60–63). Remember that a talk is always more interesting if you have some exhibits to show and pass round the class.

5. Make a list of ten solids and ten liquids which you might find in a kitchen.

Chapter 2 **Crystals**

Looking at things

In this chapter we will start thinking about the structure of matter. To do this it would seem sensible to look at it closely and perhaps the first thing would be to use a hand lens (a magnifying glass).

Experiment 2.1   Using a hand lens

Hold the lens close to your eye with one hand and bring up the object towards the lens with the other hand until you can see it clearly. Look at your finger nail, your skin, a drop of blood, a piece of material, a bit of nylon stocking. Look at the blocks you looked at earlier: blocks of iron, aluminium, wood and so on. Look at a picture from a newspaper: you may be rather surprised by what you see. Above all, look at some photographic hypo (solid), some common salt, some sugar. If possible, look at sugar of various kinds, granulated, castor, demerara and icing sugar. What do you notice about the grains of salt or the grains of sugar?

Experiment 2.2   Using a microscope

After you have used the hand lens, you might use a microscope. Basically, it consists of two lenses (or magnifying glasses) and it is more effective than a single lens. The lens nearer the object you are looking at is usually called the *objective* lens, the one nearer your eye the *eyepiece*. The metal tube between the two lenses holds these the right distance apart, the focussing knob on the side provides the fine adjustment needed to see the magnified object clearly.

The object to be looked at should be placed on the platform. To look at most objects, it is better to put them on a microscope slide (a thin strip of clear glass) which is then

Granulated sugar

Salt crystals

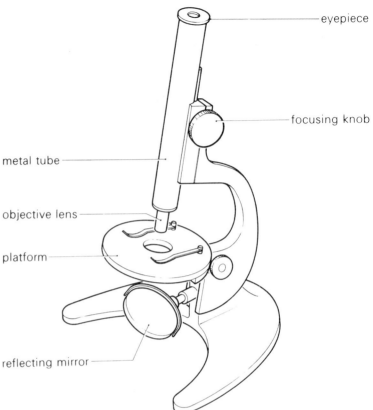

eyepiece

focusing knob

metal tube

objective lens

platform

reflecting mirror

put over the hole in the platform. For some objects it may be best to put them between two such slides. Most microscopes have a mirror under the platform. This should be tilted so that it reflects the maximum amount of light from a window or a lamp on to the object, through the hole in the platform.

Look at a drop of blood, a human hair, a dead fly, some grains of salt, some grains of different kinds of sugar. What do you notice about the salt, or about the sugar?

Regularities

The interesting thing about the salt, or the sugar, is the regularities between one grain and another. Not all the sugar grains are the same size, nor do they necessarily have the same shape, but they all appear to have the same angles between faces. To these regular structures we give the name *crystals*.

14

The photograph on the left shows a large crystal of alum. Notice the regularity of the angles between the faces.

Crystals

The photographs lower down the page show examples of crystals which occur in nature. They all show a lot of hard, smooth faces with sharp edges. These flat surfaces can be very good reflectors when light shines on them.

If you hold a piece of granite so that light can be reflected off it and if you rotate it slowly, you should notice parts of the surface suddenly appearing bright or sparkling and thereby revealing the presence of crystals*. Similar flat surfaces can be seen in the photograph on page 16 of the crystals of pyrite.

*Three main materials can be seen in a specimen of granite: shiny flaky mica, clear glass-like quartz and some felspar, usually pink or white in colour.

Alum crystal

Natural crystals

Pyrites (left) Alum model

What is the reason for this regularity? The photograph above shows a model built with polystyrene spheres. It shows the same sort of regularity as the alum crystal at the top of page 15. One possible explanation of the regularity of crystals is that they too are built up from many small 'building blocks'.

The 'building blocks' in one case are the polystyrene spheres. The 'building blocks' for the crystals might possibly be particles such as atoms or molecules. We will now see if there is any other evidence to support this idea of crystals being made of particles stacked up together.

Crystals growing

You have already looked at the regular shape of salt crystals. How do the crystals get like that? To find out, it is sensible to watch crystals growing and a number of experiments are possible.

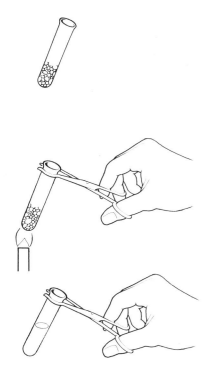

Experiment 2.3   Hypo crystals

Take a test tube and fill about a third of it with hypo crystals. If you heat it gently the crystals will dissolve. Then let the test tube slowly cool to room temperature by placing it in a test tube rack. Hold the test tube in your hand and drop into it a small hypo crystal. Something immediately starts to happen. Did you notice any change in temperature?

The experiment can easily be repeated by once again warming the crystals so that they dissolve. You can cool the test tube under a running tap, but if you try to hurry things up like this, you may find crystals suddenly forming even before you have put in a small hypo crystal.

This experiment shows crystals growing very quickly. The next experiment produces larger single crystals, but it takes much longer.

Experiment 2.4   Growing an alum crystal

The object of this experiment is to grow a single crystal of alum and it is well worth taking some trouble with it. You can do it in your school laboratory using a beaker, or you can do it at home using a jam jar.

You will need about 150 g of potassium alum. Put this in a beaker and fill it nearly full of water. Heat it and stir until the alum has dissolved. (If you do this at home with a jam jar, you can heat it in a saucepan of water on the kitchen cooker.) Pour off the warm liquid into another perfectly clean dry beaker, or jar, and leave it to cool.

To start the crystal growing you need a small crystal as a *seed* on which the crystal can form. To get a suitable seed pour off a little of the liquid into a small dish, and leave it lightly covered to keep out dust. In a day or two, a few small crystals will appear. Choose one of them on which to grow your big crystal. Dry it carefully, tie a piece of cotton round it and hang it in the original solution.*

* There is a limit to the amount of alum which can normally be dissolved in a certain volume of water at a particular temperature. When this amount is dissolved, the solution is said to be *saturated*. If less is dissolved, it is *unsaturated*. If the alum is dissolved in hot water until it is saturated and the water is then allowed to cool, the solution will be *super-saturated*. If a grain or two of alum is put in a super-saturated solution, many small crystals will form until the solution is no longer super-saturated. If a crystal is put in an unsaturated solution, it may dissolve although it will depend on the temperature whether this happens quickly or not.

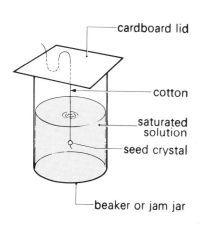

cardboard lid

cotton

saturated solution

seed crystal

beaker or jam jar

Another method of producing seed crystals is to dip a piece of thread in the saturated solution and then hang it up to dry. Small crystals will appear on it and the whole thread can then be hung in the solution. After a day or two examine the thread to see which crystals have grown, and break off all but the best.

A good way to suspend the crystal is to thread the cotton through three holes in a piece of cardboard placed over the beaker or jar. In this way it is easy to adjust the height of the crystal. Leave the crystal in the solution. As the solution slowly evaporates alum will be deposited on your seed crystal which will begin to grow.

With patience you can grow a really large crystal, but there are some conditions necessary for this.

a. Do not let the solution evaporate too quickly or you will just get a lot of small crystals around the bottom or sides of the jar and the alum will not be deposited on your crystal.

b. Keep your jar where the temperature does not change much. If your jar suddenly warmed up, the solution might become unsaturated and your crystal would dissolve instead of growing.

c. Remove any small crystals which may form on the thread: they might fall on your crystal and grow there ruining its shape.

d. If too much of the solution has evaporated and you want to continue growing your crystal, you can always take it out and put it in a fresh lot of saturated solution. It is important to remember that a good crystal will take several weeks to grow and patience is needed.

Experiment 2.5   Salt crystals

Put some common salt in a test tube, fill it half full with water. Shake it until the salt has dissolved. Add more salt and shake again. Continue the process until no more can be dissolved. You will then have a saturated salt solution.

Warm a microscope slide over a low Bunsen flame and put a drop of the salt solution on the slide. Put the slide on the platform of a microscope and then watch the crystals forming at the edge of the drop. (To get the drop in focus, it may help to put a pin on the slide and focus on that.)

saturated salt
solution
on slide

objective lens

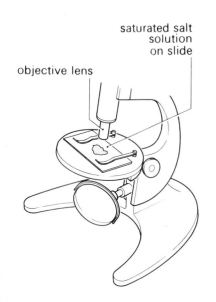

Experiment 2.6　Growing salol crystals

A very good substance for showing crystals growing is salol. It is not expensive and can be bought at chemist shops. Put a little on a microscope slide and place it on a radiator or convector heater (a temperature of 43 °C is sufficient). Remove the slide and watch what happens to the liquid when a very small seed crystal is put in it. It is better if you watch with a hand lens or use a low-powered microscope.

You will see the crystals growing. They usually start at the edge of the liquid. When they grow they always have a particular shape. After a while they 'bump' into each other and growth gets restricted in that direction.

Other experiments

Instead of growing an alum crystal (as in Experiment 2.4) you may prefer to grow a copper sulphate crystal which has a beautiful blue colour. You should do the experiment in exactly the same way as described in Experiment 2.4, except that you will dissolve copper sulphate and not potassium alum to get your saturated solution.

You might also like to watch copper sulphate crystals growing on a microscope slide (although you should be warned that it will not happen as quickly as occurred with salol in Experiment 2.6). The photograph on the left shows a thin layer of such crystals on a microscope slide.

Notice how the angles between faces are always the same.

Copper sulphate crystals
Beads in a tray

Models of crystal growth

How do these experiments fit in with our idea that a crystal is made up of particles acting as building blocks?

If you take a lot of glass beads (all the same size) and pour them into a small tray so that they are one layer thick, you will notice that they settle into a definite pattern without your having to arrange them.

Of course there are some irregularities, but you will notice that there is very often a hexagonal packing, one bead with six beads all round it.

Experiment 2.7 Crystal models

Arrange several polystyrene spheres in the hexagonal pattern as shown in the margin. (You can use marbles instead, provided they are all the same size, or flat pennies.) Then add more spheres to the pattern so that it gets bigger. What do you notice about the angles as the shape grows? What do you notice about the way the other spheres pack together?

The spheres need not be arranged in this hexagonal way. They might be arranged in squares. The drawing below shows nine spheres arranged in this way on a baseboard (the baseboard merely stops them from rolling about over the bench). What would happen if some more spheres were added on top of this?

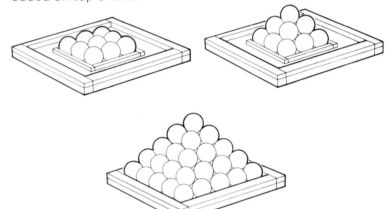

Clearly we get a pyramid. That pyramid had a base of nine balls, three in each side. The model could be extended so that there were four balls in each side of the base, or five balls as in the third drawing. And do you notice that on the side of that large pyramid the hexagonal packing appears again? Each sphere seems to be surrounded by six spheres. If you were to stick the balls together and go on adding more and more, you could build up the model shown below. This has considerable similarities with the large crystal of alum at the side. Perhaps this is good evidence that crystals are made of basic building blocks.

Cleaving crystals

Further evidence comes from the way some crystals can be cleaved. These crystals tend to split along certain flat surfaces. (Cleave is another word for split.) If the regular shape of crystals means that they are made up of layers of particles, we might expect them to cleave along certain planes or flat surfaces and this is just what happens. The photograph below shows a calcite crystal being cleaved and the second photograph shows the same thing in our model.

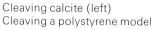
Cleaving calcite (left)
Cleaving a polystyrene model

Questions for class discussion
1. If you look at a particularly fine type of icing sugar with your microscope and you cannot see crystals, can you assume from this that the icing sugar is not crystalline?

2. You saw calcite being cleaved in your laboratory. It was very important to cleave it at the correct angle. What would happen if the angle were wrong? What would happen in your polystyrene sphere model if you tried to cleave it at the wrong angle?

3. If you take a cube of sugar and try to cleave it, you cannot do so : it shatters whatever angle is chosen. When you looked at sugar with a hand lens, each small piece looked like a crystal. Do these two things contradict each other?

Galvanised can (left)
Snow flakes

Crystals in a cast metal bar

4. You have seen hypo (sodium thiosulphate) crystals grow-ing quickly and an alum or copper sulphate crystal growing slowly. Why do you think there is this time difference?

5. The photograph on the left above shows a piece of galvanised iron. What do you notice about it? How can you explain what you see?

6. The photograph on the right above shows snow flakes. Why do they show such patterns?

Some things to try at home

1. Try to find an old brass door knob. It may have been 'etched' by the sweat of people's hands. If so, what do you notice about it?

2. Try to find a piece of broken metal and examine the broken part for crystals. Look for evidence of crystals in galvanised buckets or inside fruit tins.

3. Try growing a crystal of alum or copper sulphate at home. Perhaps the most important thing is to keep the jar where the temperature does not change too much: a basement is probably the best place.

Mica

4. Make saturated solutions of alum, sugar, table salt, washing soda and Epsom salts. In each case pour the saturated solution into a small glass and leave it on a shelf for a week or two. Have a look at each of them every day as they evaporate and describe what happens.

5. See if you can get a piece of mica. It cleaves very easily indeed in one direction, so easily that it almost appears to be made of a pile of very thin plates. Take a piece and cleave it into thinner and thinner slices using a one-sided razor blade.

Growing a chemical garden

Either at school or at home, you might like to have some fun growing a chemical garden. Get some water-glass from a chemist (it is sometimes used for preserving eggs) and mix it with about the same quantity of water. Put a layer of sand about 1 cm thick at the bottom of a large jam jar. Then pour the water-glass mixture carefully into the jar without disturbing the sand. Drop in some crystals of hypo (sodium thiosulphate), copper sulphate, magnesium sulphate, cobalt chloride, iron sulphate and potassium dichromate, or any other crystals your teacher can give you. Cover up the jar and leave it undisturbed and see what happens.

Homework assignments

1. Read the topic book *Crystals*, pages 13–21, and prepare a short talk on gemstones. It might be worth trying to see if a local jeweller has any imperfect stones he would lend to illustrate your talk.

2. Find out what is meant by etching and read what is said about it in the topic book *Crystals*. What can some etched metal tell you?

Conclusion

All the evidence you have seen suggests that a reasonable way to make a model of a solid is to use building blocks of small particles. We might suppose the particles are atoms, and we will assume they are until we get any evidence to the contrary. Of course we do not know why the particles hold together; we shall have to wait until a later stage before we find out.

Chapter 3 # Measurement

It is easy to make guesses about matter and its structure, but it is only possible to make real scientific progress when measurements are made. In no subject is measurement more important than in physics and it is now time to start thinking about it.

In Britain we have in the past measured road distances in miles, lengths of material in yards, sugar in pounds and beer in pints. However, the metric system is being used increasingly in our everyday life and it has for a long time been the system used in scientific work. We will use it throughout this book.

Measurement of length

The standard unit for measuring length is the metre. Your school laboratory will have some metre rules and these will help you to learn to judge lengths in metres.

The metre is too small a unit for measuring distances between towns and it is usual to use kilometres (1 kilometre = 1 000 metres). For small distances centimetres and millimetres are both used: 100 centimetres = 1 metre and 1 000 millimetres = 1 metre. There are standard abbreviations for these units and it is wise to get familiar with them.

* It is now accepted scientific usage not to put a full stop after m, cm, mm even though they are abbreviations, unless of course it comes at the end of the sentence. Another accepted usage is to omit the traditional commas in writing, say, 1,000,000, but instead to write it 1 000 000. These conventions will be followed in this book.

Unit	Abbreviation*
metre	m
kilometre	km
centimetre	cm
millmetre	mm

$$1 \text{ km} = 1\,000 \text{ m}$$
$$100 \text{ cm} = 1 \text{ m}$$
$$1\,000 \text{ mm} = 1 \text{ m}$$

$$1 \text{ m} = \tfrac{1}{1000} \text{ km} = 0.001 \text{ km}$$
$$1 \text{ cm} = \tfrac{1}{100} \text{ m} = 0.01 \text{ m}$$
$$1 \text{ mm} = \tfrac{1}{1000} \text{ m} = 0.001 \text{ m}$$

You will come across kilo-, centi- and milli- a lot in the future. 'kilo' means 1 000 times, 'centi' means $\tfrac{1}{100}$th and 'milli' $\tfrac{1}{1000}$th.

Quick exercises

1. Measure the length of this page in centimetres. Write down the length in metres and in millimetres.
2. Measure the thickness of this book in centimetres.
3. Measure your own height in centimetres and in metres.
4. Compare the length of a metre rule with one yard. Which is bigger? Is a 100 metre race longer or shorter than a 100 yard race? By how much would they differ?

Powers of ten

The distance from the Earth to the Sun is about 150 000 000 000 m. This involves a lot of noughts and scientists do not find it very convenient writing numbers like that. They prefer to use a shorthand way of writing it. 150 000 000 000 is 1.5 multiplied by 10 a number of times. How many times? How many places would the decimal point have to be moved to change 1.5 into 150 000 000 000? The answer is 11 so we can write 150 000 000 000 as 1.5×10^{11}, and that is much quicker to do.

Similarly
$$50 = 5 \times 10 = 5 \times 10^1$$
$$500 = 5 \times 10 \times 10 = 5 \times 10^2$$
$$5\,000 = 5 \times 10 \times 10 \times 10 = 5 \times 10^3$$
$$5\,000\,000 = 5 \times 10 \times 10 \times 10 \times 10 \times 10 \times 10 = 5 \times 10^6$$

You will study powers of ten in more detail in your mathematics course. We will, however, use it as a convenient shorthand in this book.

Standard form
The distance from the Earth to the Sun can be written in the following ways:

$150 \times 10^9\,\text{m}$ $15 \times 10^{10}\,\text{m}$ $1.5 \times 10^{11}\,\text{m}$ $0.15 \times 10^{12}\,\text{m}$

All of these are equivalent mathematically but scientists usually prefer to write such a quantity with one figure in front of the decimal point. The standard form would be $1.5 \times 10^{11}\,\text{m}$. It is useful to get into the habit of writing numbers like this.

Measurement of area and volume

You have learnt in your mathematics how to find the area of a rectangle: you multiply the length by the breadth. Similarly you have also learnt how to find the volume of a rectangular block: you multiply the length by the breadth by the height.

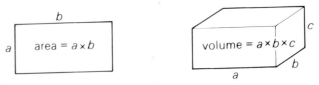

The units for measuring area and volume are also based on the metre: square metres for area, cubic metres for volume. There is a very convenient shorthand for writing square metres as m^2, and cubic metres as m^3 and these will be used in this book. Similarly cubic centimetres can be written as cm^3.

Suppose the drawing opposite represents 1 square centimetre (not drawn to scale). As there are 10 millimetres in 1 centimetre, there will be 10 millimetres along each side. Each little square in the drawing is 1 square millimetre. How many little squares are there? It follows that

$$100\,\text{mm}^2 = 1\,\text{cm}^2$$

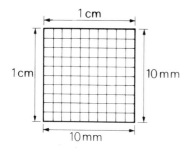

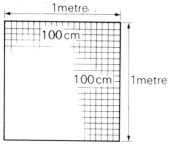

Since there are 100 cm in 1 m, there will be 100 cm along each side of a 1 m square. Therefore there will be 100×100 small centimetre squares in 1 square metre. In other words
$$10\,000 \text{ cm}^2 = 1 \text{ m}^2$$
Similarly, there will be $10 \times 10 \times 10$ cubic millimetres in 1 cubic centimetre. In a cubic metre, there will be $100 \times 100 \times 100$ centimetre cubes. In other words
$$1\,000\,000 \text{ cm}^3 = 1 \text{ m}^3$$
Using our power-of-ten notation, we can write
$$10^4 \text{cm}^2 = 1 \text{ m}^2 \quad \text{and} \quad 10^6 \text{ cm}^3 = 1 \text{ m}^3$$

Questions for homework or class discussion

1. What is 10×10 in words?
Write it down as a power of 10.
Do the same for $10 \times 10 \times 10$.
And for $10 \times 10 \times 10 \times 10 \times 10 \times 10$.

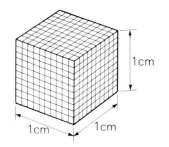

2. The mass of the Earth is about
$$5\,980\,000\,000\,000\,000\,000\,000\,000 \text{ kg}.$$
Write this down using powers of ten. You may have written your answer as something $\times 10^{22}$, or as something $\times 10^{23}$, or as something $\times 10^{24}$. Now write it down in all three ways. Which of these three ways is the standard form preferred by scientists?

3. What is the area of a rectangular field 220 m by 50 m?

4. What is the area of a rectangular sheet of metal 3 m long and 20 cm wide? Give the answer in cm².

5. What is the volume of a block of metal 5 cm long, 4 cm wide and 3 cm high?

6. What is the volume of a metal cube with side 4 cm?

7. A wooden crate is 2 m long, 1 m wide and 1 m deep. How many 1 cm cubes could you fit into the crate? Express your answer in powers of 10.

8. This is a map of an island. The lines drawn on the map are each 1 km apart. Estimate the area of the island in square kilometres. (*Hint:* what is the area of one square, how many complete shaded squares are there, how many squares have more than half shaded, how many have less than half shaded?)

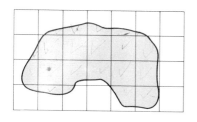

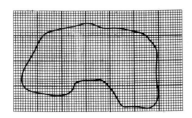

Measuring irregular areas

You can find the area of a rectangular area by measuring the length and the breadth, and multiplying the two together. Question 8 above suggests how you might estimate the area of an irregular shape. But suppose you wanted to know the area with rather greater accuracy. What could you do?

You might divide the big squares into a hundred little squares, and then you could count first the number of complete big squares and then the number of little squares. Each big square will represent 1 km², each little square will be $\frac{1}{100}$ km². This will give you a more precise answer.

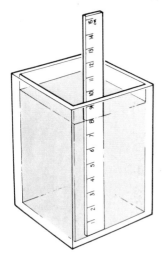

Measuring the volume of liquids

It was easy to find the volume of a solid rectangular block but what can we do about liquids? They do not have a definite length, breadth and height, but take the shape of the container in which they are put. But that gives us the clue how to measure their volume.

Pour the liquid, whose volume you want to know, into a perspex container. (Those recommended for school use have *internal* length and breadth of 5 cm.) Measure the length, breadth and height, and multiply all three together, and you have the volume.

Experiment 3.1   Checking the scale on a measuring cylinder

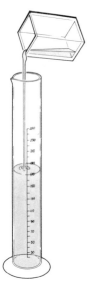

A very convenient way of measuring the volume of liquids is to use a measuring cylinder. These have a scale marked on them. To check whether these scales are correct, put water in one of the perspex containers to a depth of 4 cm. The volume of water will be $5 \times 5 \times 4$ cm³ or 100 cm³. Pour this water into the measuring cylinder and see if it comes to the 100 cm³ mark. How would you check the 200 cm³ mark? And the 250 cm³ mark on a 250 cm³ measuring cylinder?

Experiment 3.2 Measuring irregular solids

A rectangular block is called a regular solid and it is easy to calculate its volume. But many solids are irregular — a rock, a pair of scissors, a bicycle — and their volumes cannot be easily calculated from a few measurements.

One way to find the volume of a rock is shown below. Put water in a measuring cylinder so that it is about half full and read the volume. Then lower the rock on a piece of cotton into the water. The level will rise and the new volume is read. The difference between the two readings gives the volume of the rock.

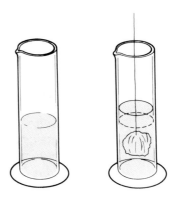

Accuracy

Suppose you measure the height of a book and you say it is 21 cm. What does this mean? To a physicist, it means that the length is neither 20 cm nor 22 cm, but that it lies between 20.5 cm and 21.5 cm. That sort of accuracy is sufficient if you are deciding whether the book will fit on bookshelves 25 cm apart. But it would not be sufficiently accurate for the length of a piece of wood to be used for a chair leg: to that accuracy the four legs might be very different. One leg may be 20.6 cm, another 20.8 cm, the third 21.2 cm and the fourth 21.4 cm, and you would accidentally have made a rocking chair.

To describe something more accurately it is necessary to give further decimal places:

21.5 cm	means it is between	21.45 cm	and	21.55 cm
21.55 cm	means it is between	21.545 cm	and	21.555 cm
21.555 cm	means it is between	21.5545 cm	and	21.5555 cm

Similarly, 21.0 cm means the length lies between 20.95 cm and 21.05 cm.

It does not have any meaning to say that the height is *exactly* 21 cm, unless you mean 21 followed by an infinite number of noughts and that, of course, is absurd.

Averages

Whenever possible a scientist does not like to rely on only one reading when he is making a measurement. He prefers to take several readings and calculate the average. To do this he adds up the readings and then divides by the number of readings.

It can sometimes happen that a bad mistake is made in one of the readings, in which case that reading should be omitted when working out the average, as will be shown in one of the examples below.

Questions for homework or class discussion

9. If you buy 8 m of rope, it is impossible to cut it at *exactly* 8 m. Strictly speaking, 8 m would mean anything between 7.5 and 8.5 m although the salesman is usually careful to make it over 8 m rather than under. Why does he do this? If you want it between 7.95 m and 8.05 m, how should you write the length? If you want it between 7.995 m and 8.005 m, how should you write it?

10. A boy measures the length of a table and he says it is 1.231 42 m. If he measured it with a tape measure marked only in centimetres, is this a reasonable answer? What sort of accuracy could you expect with such a tape measure? If he had measured it with a tape measure marked in millimetres, write down what would have been a reasonable answer for him to give. Why is a fabric tape measure not a good thing to use for scientific measurements?

11. A girl times herself on each occasion when she walks round a field. The readings are 11 minutes, 12 minutes, 13 minutes, 12 minutes. What is the average time?

12. A boy used a watch with a seconds hand to time a ball rolling down a slope. He took three readings. They are 12 seconds, 12 seconds, 13 seconds. He adds them up and gets 37 seconds. He divides by 3 and says the average is 12.333 3 seconds. Is this a sensible answer?

13. Six boys are asked to measure the length of a piece of wood using a metre rule. The lengths they write down are the following:

63.2 cm	63.9 cm
63.8 cm	63.9 cm
36.4 cm	63.1 cm

One of the readings is a bit strange. Which one? What do you think that boy did wrong? Calculate an average value for the length of the piece of wood.

Estimations

Scientists are good at making estimates. An estimate may be only a quick rough guess, but it is not necessarily bad science because of that. Suppose you have a doorway 1 m wide and you want to know if a table will go through it. A rough estimate that the table is 80 cm wide is quite sufficient to tell you what you want to know, even though it may in fact be 75 cm or 84 cm.

Rough estimates also help us to avoid careless mistakes in arithmetic: they are a valuable check to our work. Suppose a man wants to know the size of the floor of a large room. He measures the length as 12.4 m and the breadth as 9.3 m. He multiplies the two together and writes down the area as 1 153.2 m². If he does a quick estimate taking the length as 10 m and the width as 10 m, he will realise the answer should be about 100 m² and he quickly sees his mistake in the position of the decimal point.

Making estimates may seem difficult at first, but you will find that it becomes much easier with a little practice. Making estimates can be rather fun once you realise that no one is looking for an exact or *correct* answer.

Suppose you want to estimate the contents of a box of drawing pins. At first you may think you have no idea. But is it one? No, many more than that. Ten? No, more than that. Is it 1 000? No, that is far too many. Five hundred perhaps? Less than that. Already, you have estimated that it is greater than 10 and less than 500. Perhaps 100 might be a sensible guess. But, you might say, it depends on whether the packet is large or small. All right, if it is small we will change the estimate to forty and if it is large to 200. These are not at all likely to be exact answers to the question, but they are not bad guesses. You might want to estimate the number of words in this book. It is obviously more than 100 and more than 1 000. Could it be a million or a thousand million? You have no idea. All right, how many pages are there? About 200. Good, how many words on a page? The trouble is there are a lot of pictures on each page. All right, what is the average number of lines on a page? You might count the number of lines on four or five pages in the middle of the book. Perhaps the average is about forty. And how many words to a line? An estimate might be about ten. That means about 400 words to a page and, with 200 pages, there must be about 80 000 words. To find the exact answer would take someone a very long time to count and would it really make any difference if you knew it was 83 475? Making sensible estimates is good science.

Questions for homework or class discussion

14. Estimate the height of the room in which you are sitting. It may help if you start by thinking of the height of a man. Can one man stand up in the room? Could one man stand on the shoulders of another man and still be upright? Could more men stand on each other's shoulders? Give your answer in metres.

Estimate the length and breadth of the room in metres. Calculate the area of the room. Give your answer in square metres.

Use your estimates of the length, breadth and height of the room to calculate its volume. Give your answer in cubic metres.

15. Estimate the length of a car in metres.

Estimate the number of cars, bumper to bumper, which could be put into 100 metres.

Estimate the width of a car in metres. How many cars could be put side by side in 50 metres?

Use this to decide the maximum number of cars that could be fitted into a field 100 metres by 50 metres.

In a car park it would not be possible to put cars as close as that. Estimate what you think would be the maximum number of cars you could conveniently park in such a field.

16. Estimate the diameter of an apple in centimetres.

Apples are usually sent to market in boxes. Estimate the length, breadth and height of a conveniently sized box. How many apples could be put along one side of the box? How many apples could you fit in the bottom layer? Estimate the total number of apples you could get into the box.

17. Estimate the size of a grain of salt. How many grains would there be in a pinch of salt? Estimate the number of grains there might be in a packet of table salt bought at a grocer's shop.

Chapter 4 **Mass and density**

A young child discovers that unless a ball or a brick is held up it will fall to the ground. As scientists, we explain this by means of gravity. It was Isaac Newton who developed this idea of gravity and you have probably heard the story — which may or may not be true — that he watched an apple fall and realised that it was the gravitational force of the Earth pulling on the apple which made it fall.

Scientists speak about this *force* due to gravity acting on a brick as the *weight* of the brick. But the force of gravity is a little bit less up a mountain than it is at sea level and this means that higher up the mountain the weight would be less. You also know that gravity is a lot less on the surface of the Moon than it is on the Earth, so that the weight of a brick on the Moon will be less than it is on the Earth. On the other hand there is just as much matter in the brick on the Moon as there is when it is on the Earth. Scientists, therefore, prefer to talk about the *mass* of the brick and this measures the amount of matter there is in it. We measure this mass in kilograms (kg). For scientific purposes we will always use kilograms. A kilogram of butter has exactly the same mass of 1 kg whether it is measured on the Moon or at the top or bottom of a mountain on the Earth. There is the same amount of butter in each case. But whereas 1 kg of butter is an unchanging mass, it will have different weights on the Moon and on the Earth because weight depends on gravity.

The kilogram is sometimes too large a unit and the gram is used instead. 1 000 grams is the same as 1 kilogram. It is sometimes convenient to use a milligram (abbreviated to mg), which is one thousandth of a gram.

Unit	*Abbreviation*
kilogram	kg
gram	g
milligram	mg

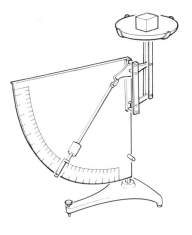

$$1\,000\text{ g} = 1\text{ kg}$$
$$1\text{ g} = \tfrac{1}{1000}\text{ kg} = 0.001\text{ kg}$$
$$1\text{ mg} = \tfrac{1}{1000}\text{ g} = 0.001\text{ g}$$

It is very unfortunate that, in everyday use, people often use the word *weight* when they mean the word *mass*. For example, you may find a packet of butter which is marked 'net weight 250 g' when it ought to say 'net mass 250 g'. It is also unfortunate that people use the word *weighing* when they mean *finding the mass* of something. It is wise to get into good habits and in this chapter we will do our best to avoid the word weight and to refer correctly to the mass of an object.

with mass here

read this scale

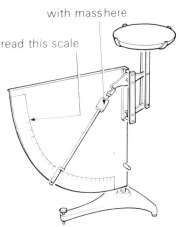

Measuring mass

One of the easiest ways to find the mass of an object is to put it on a lever-arm balance, sometimes called a Butchart balance.

Experiment 4.1   Measuring mass
There is usually a screw on the base of the balance and this should be adjusted before use so that the reading is 0 when there is nothing on the pan.

Put the block to be measured on the pan and read off the mass against the pointer on the scale.

Most lever-arm balances have two scales. If the movable mass is in the upper position (as in the first drawing) read the upper scale. If it is in the lower position (as in the second drawing) read the lower scale.

With some balances you can change your reading by moving your head. This may cause an error, called a *parallax* error, which can be a bit of a nuisance. The best way to keep this error to a minimum is always to take readings with your eye directly in front of the pointer and the scale.

A good scientist would probably want to check the accuracy of his instrument. How could you do this? (*Hint:* There will be some 100 g and some 1 kg masses in your laboratory which might help you.)

Take some blocks, all the same size, of the following:

with mass here

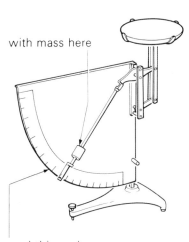

read this scale

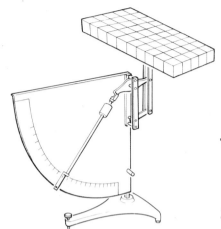

aluminium, iron, softwood, hardwood, paraffin wax, expanded polystyrene. Find the mass of each and write down the answers in your notebook.

Finding the mass of the expanded polystyrene block is a bit difficult. How could you get over the difficulty? Perhaps the drawing on the left gives a clue.

Questions for class discussion

1. As a result of the previous experiment, arrange in order of increasing mass the blocks of aluminium, iron, softwood, hardwood, paraffin wax and expanded polystyrene. Does everyone in the class agree with the order?

2. 'Iron is heavier than expanded polystyrene'. Is this a sensible statement?

3. Which has the greater mass, a kilogram of iron or a kilogram of expanded polystyrene?

4. A tonne is equal to 1 000 kilograms (it is very roughly equivalent to the English ton). Which would be easier to store: a tonne of iron or a tonne of expanded polystyrene?

Density

We have to be very careful before we make remarks like 'iron is heavier than aluminium', because a large block of aluminium may well have more mass than a small block of iron. How can we compare different substances? This can be done by finding the mass of the *same volume* of each substance.

What volume shall we take? A convenient volume would be 1 cubic centimetre. (We will see later that scientists often prefer to take the volume as 1 cubic metre, but we will start with 1 cubic centimetre.)

Look back in your notebook to see what was the mass of your aluminium block. Write down the length, breadth and height of the block. How many cubic centimetres are there in it? What is the mass of 1 cubic centimetre? This quantity, the mass of 1 cm³ of the substance, is called the *density* of the substance. It will be measured in grams per cubic centimetre (i.e. g/cm³).

For example, a block which measures $3\,\text{cm} \times 4\,\text{cm} \times 5\,\text{cm}$ has a volume of $60\,\text{cm}^3$.

If $60\,\text{cm}^3$ of the substance has a mass of $120\,\text{g}$ then $1\,\text{cm}^3$ has a mass of $120\,\text{g} \div 60 = 2\,\text{g}$.

Thus the density of the block is $2\,\text{g/cm}^3$.

Blocks of material

Experiment 4.2   Measuring densities

Measure the masses of various blocks of different materials using the lever-arm balance. Draw a table similar to the one below and write down the masses in the second column.

Material	Mass g	Length cm	Breadth cm	Height cm	Volume cm³	Density g/cm³
aluminium iron softwood hardwood lead glass marble perspex paraffin wax						

Measure each block using a plastic measuring rule and add the details to the table. Then calculate the volume, putting the value in the next column. Finally, work out the density. Remember that

$$\text{DENSITY} = \frac{\text{MASS}}{\text{VOLUME}}$$

Make a list of the blocks you have measured, arranging them in order of increasing density.

Questions on density

1. A block of iron is 5 cm × 5 cm × 4 cm. It has a mass of 750 g. What is the volume of the block? What is its density?

2. If the density of lead is 11 g/cm³, what is the mass of (a) 1 cm³, (b) 10 cm³, (c) 100 cm³, of lead?

3. The density of balsa wood is 0.2 g/cm³. What is the mass of (a) 1 cm³, (b) 5 cm³, (c) 10 cm³, (d) 50 cm³, of balsa wood?

4. The density of a substance is 3 g/cm³. What is the volume of a piece of the substance which has mass of (a) 27 g, (b) 270 g?

5. Water has a density of 1 g/cm³. Ice has a density of 0.9 g/cm³. Which has the bigger volume: 10 g of water or 10 g of ice?

6. Write down your own mass in kilograms. What is your mass in grams? The density of the human body is about the same as that of water. Use this fact to calculate the volume of your body.

7. The density of lead is 11 g/cm³, the density of iron is 7.5 g/cm³. Which has the greater mass: 16 cm³ of iron or 10 cm³ of lead?

8. (Difficult) The density of gold is 19 g/cm³. A block of gold is measured and the volume is found to be 10 cm³. It is found that the mass of the gold is only 120 g. Can you suggest a possible explanation of this?

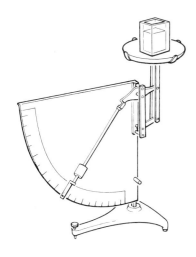

Measuring the density of a liquid

If you want to find the mass of some liquid, you might perhaps pour it into the pan of a lever-arm balance. But you will not be very popular in your laboratory if you do this.

To find the mass of a liquid, first find the mass of an empty container. Then put the liquid in the container and find the combined mass. Take one from the other and you are left with the mass of the liquid.

Experiment 4.3   Measuring the density of a liquid

Take a rectangular perspex container. Find its mass when it is empty. Put the liquid in the container to a depth of, say, 4 cm. If the inside dimensions of the container are 5 cm × 5 cm, what is the volume of the liquid?

Find the combined mass of the container and liquid. Subtract the mass of the container in order to find the mass of the liquid.

Knowing the mass of the liquid and its volume, the density can be calculated.

Use this method to find the density of a number of different liquids.

Floating and sinking in liquids

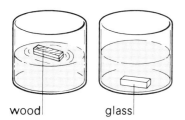

wood   glass

Some solids will float when placed in a liquid, others will sink.

If a dish is half-filled with water as shown, what happens when a block of softwood is placed in the water? What happens when a similar shaped glass block is placed in it?

The density of water is 1 g/cm³. The density of softwood is about 0.5 g/cm³, but the density of glass is about 2.5 g/cm³. It looks as though the solid floats when its density is less than that of the liquid, but sinks when it is greater. We can test this theory with other liquids.

Mercury has a density of 13.6 g/cm³. A glass block would

therefore float in it. A lead block with density 11.4 g/cm³ would also float in it, but a block of gold (density 19.3 g/cm³) would not.

Questions for class discussion

1. An iceberg will float in the sea with some of it above the surface (though a lot more below the surface). What can this tell us about the density of ice compared with the density of water?

2. When a boy floats in water, there is not much of him above the surface. What does this tell us about the density of the boy?

3. If the same boy floats in the Dead Sea, he will lie with rather more of his body out of the water. What does this tell us about the water in the Dead Sea?

4. Three liquids were poured into a glass jar, shaken up and left to stand. When they had settled the appearance was as shown in the diagram. Which liquid, A, B, or C, has the greatest density? Which liquid has the least density? If A is water and C is mercury, why cannot B be a solution of copper sulphate?

5. If a balloon is filled with hydrogen and let go, it rises upwards. But if a balloon is filled with carbon dioxide, it falls downwards. Why is this?

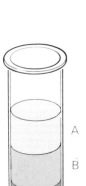

Air

We have found the density of solid substances and of liquids, but what about gases? How could we measure the density of air?

We speak of a bottle as being half-full of water. Does that mean that it is also half-empty? No, because there is air in the other half. In the exhibition of materials in your laboratory at the beginning of this course there were two bottles, both of which looked empty, but one was labelled 'air' and the other 'vacuum'. How do we know that the second bottle had nothing inside it? You cannot open it and look inside. Some demonstration experiments will help to answer this.

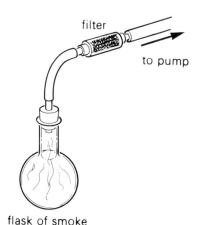

filter

to pump

flask of smoke

Demonstrations with a vacuum pump

1. *To show the effect of a pump.* Put some smoke in a round flask. Connect the flask to a pump through a filter and watch the smoky air being pumped out.

2. *To show the effect of air pressure.* Connect the pump to a large polythene bottle or can. Watch the can collapse as the air is pumped out.

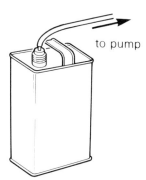

to pump

3. *To investigate the bottles labelled 'air' and 'vacuum'.* The two bottles should be similar and one is attached to the vacuum pump so that the air is drawn out. The clip on the tubing is closed before it is disconnected from the pump. Open each bottle in turn *under water* in a large transparent trough and see what happens. How can you explain the difference?

Experiment 4.4 First attempt to find the mass of a sample of air

When finding the mass of a liquid, you measured the mass of an empty container and then the same container with the liquid in it. Does the above demonstration with a bottle labelled 'vacuum' suggest how you might find the mass of a sample of air?

The drawing below shows what you might do. First adjust the balance. Find the mass of the bottle with air in it. Then use a pump to take the air out of the bottle. Then find the new mass. What is the change in mass?

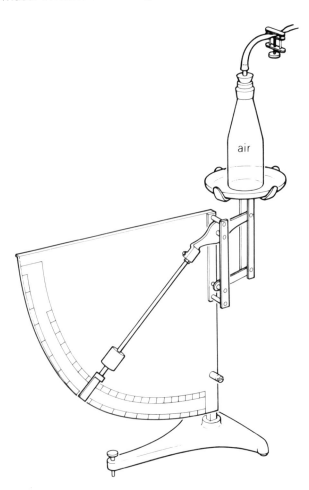

Experiment 4.5   Second attempt to find the density of air

The first attempt to find the mass of air failed. This was because the mass of the air taken out of the bottle was too small and the change was too slight. How could we get a bigger change? Perhaps use a much bigger bottle. But that could not be used on the balance. What else might we try?

Instead of using a bottle, use a large plastic container as shown in the diagram. Instead of taking air out of it, force in some extra air using a foot-pump. The object of this experiment is to find the mass and the volume of the extra air that has been added.

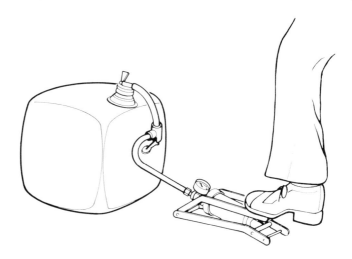

First find the mass of the container with air in it at ordinary pressure. Then connect the foot pump and force in extra air. Close the tap on the container and find the new mass. From this you can calculate the mass of extra air added.

This extra air has been squeezed into the container. The next problem is to find what would be the normal volume of this air (that is, the volume at ordinary pressure). The way to do this is to use a large trough of water in which is placed a perspex box 10 cm × 10 cm × 11 cm. The box is immersed full of water in the trough with the open side downwards.

A rubber tube from the container is put in the water with its end under the box. Open the tap so that air enters the

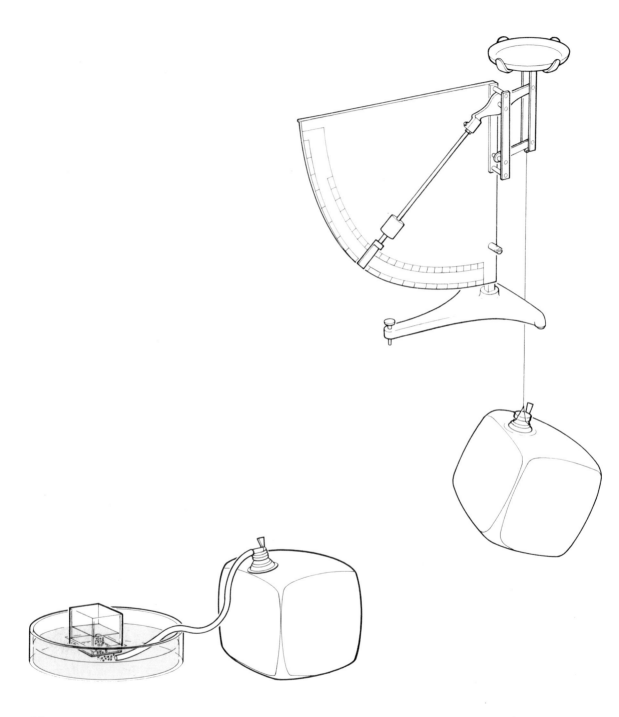

inverted box. Close the tap when the air gets to the 10 cm mark. The air is released and the box is refilled with water. Repeat the process seven or eight times until no more excess air comes out. Each time 1 000 cm³ of air is let out, so you will find out the volume of the extra air. But

$$\text{DENSITY} = \frac{\text{MASS}}{\text{VOLUME}}$$

and as you know the mass and the volume, you can calculate the density.

The first experiment showed that measuring the mass of a sample of air was difficult. It is therefore a great achievement that you have managed to get a figure for the density of air. Under normal conditions, the density of air can be taken to be 0.001 2 g/cm³.

Questions for class discussion

1. Estimate the length, breadth and height of the room in which you are sitting. What is the volume of the room, in cubic metres and in cubic centimetres?

2. What is the mass of air in the room?

3. What volume of water would have the same mass?

Values for the densities of different substances

In this chapter we have measured our masses in grams and our volumes in cubic centimetres because it was convenient to do so. But scientists prefer to use kilograms for their standard unit of mass and cubic metres instead of cubic centimetres. Sometimes you will find the densities measured in grams per cubic centimetre, sometimes in kilograms per cubic metre. We shall give the values in both units below.

There are 10^6 cm³ in 1 m³,

but 1 cm³ of water has a mass of 1 g,

so 1 m³ of water (i.e. 10^6 cm³) has a mass of 10^6 g or 10^3 kg

Thus 1 m³ of water has a mass of 1000 kg.

From this it follows that the density of water is

1 g/cm³ or 1000 kg/m³.

Other values are given in the table below.

Substance	Density in g/cm³	Density in kg/m³
aluminium	2.70	2 700 or 2.70 × 10³
copper	8.94	8 940 or 8.94 × 10³
gold	19.32	19 320 or 19.32 × 10³
iron	7.86	7 860 or 7.86 × 10³
lead	11.35	11 350 or 11.35 × 10³
nickel	8.90	8 900 or 8.90 × 10³
platinum	21.45	21 450 or 21.45 × 10³
silver	10.50	10 500 or 10.50 × 10³
zinc	7.13	7 130 or 7.13 × 10³
expanded polystyrene	0.015	15
glass	2.8	2 800
marble	3.2	3 200
paraffin wax	0.95	950
perspex	1.2	1 200
softwood	0.5	500
hardwood	0.7	700
methylated spirits	0.79	790
mercury	13.55	13 550
turpentine	0.86	860
water	1.00	1 000
air	0.0012	1.2

Chapter 5 **Revision tests A**

Revision test A1

1. Write down $3 \times 10 \times 10 \times 10 \times 10$ using powers of ten.
2. The mass of the Earth is about
$$6\ 000\ 000\ 000\ 000\ 000\ 000\ 000\ 000\ 000 \text{ kg.}$$
Write this down using powers of ten.
3. How many grams are there in a kilogram?
4. Write down the mass of the Earth in grams using the powers of ten notation.
5. $3\ 200 \times 10^5$ is the same as 32 times what power of ten?
6. $3\ 200 \times 10^5$ is the same as 3.2 times what power of ten?
7. Express 0.3 metres as a fraction of a metre.
8. Express 0.03 m as a fraction of a metre.
9. Express 60 cm as a fraction of a metre.
10. Express 5 mm as a fraction of a metre.

Revision test A2

1. How many millimetres are there in 1 centimetre?
2. How many millimetres in 5 cm?
3. How many centimetres in 1 m?
4. How many metres in 1 km?
5. How many centimetres in 1 km?
6. How many centimetres in 10 m?
7. How many centimetres in 0.1 m?
8. How many millimetres in 0.1 m?
9. How many millimetres in 0.01 m?
10. How many millimetres in 0.001 m?

Revision test A3
What is the area of a rectangular space:

1. 5 m long and 3 m wide?

2. 5 m long and 30 m wide?

3. 50 m by 30 m?

4. 5 m by 3 cm? (Answer in cm^2 and in m^2.)

5. 5 cm by 3 mm? (Answer in cm^2 and in m^2.)

What is the volume of a rectangular block of metal:

6. 4 cm long, 3 cm wide, 2 cm high?

7. 5 cm long, 5 cm wide, 4 cm high? $100cm^3$

8. 5 cm long, 5 cm wide, 4 mm thick? (Answer in cm^3.)

9. 5 m long, 2 cm wide, 2 cm thick? (Answer in cm^3 and in m^3.)

10. How many 1 cm cubes would fit into a crate 2 m long, 2 m wide, 1 m high?

More estimates A4
Estimate the following:

1. The number of metres between the entrance to the school and your laboratory.

2. The number of saucepans sold in England in a year.

3. The number of books in your school library.

4. The number of loaves of bread eaten by a family of two adults and two children in a year.

5. The number of times the ball is kicked in a game of football.

6. The number of grains of salt in a teaspoonful of salt.

7. The number of sheets of paper (including pages in exercise books) used by an 11-year-old schoolboy in a year.

8. The number of words in St Mark's Gospel.

9. The number of leaves on an oak tree.

10. The number of lamp-posts in your town (or village).

Descriptive revision test A5

1. If you had a small crystal of copper sulphate, how would you grow a larger one?

2. You have probably grown some crystals at your school. Write a set of instructions to a friend at another school telling him how to grow crystals. Assume he is not a very good scientist, so give him plenty of details, including any warnings you think necessary about precautions to take in order to get good crystals.

3. You will have seen crystals growing under a microscope. What do you notice about the shapes in which they grow? How do you explain this growth using a model of matter made of small particles?

4. Describe an experiment you would do to find the volume of an irregular object like a screw-driver.

5. A measuring cylinder arrives at your laboratory, but the manufacturer has forgotten to mark any figures on it. Describe an experiment you could do in order to put it right.

Revision test A6

1. A block of metal is $5 \text{ cm} \times 4 \text{ cm} \times 3 \text{ cm}$. It has a mass of 240 g. What is its density in g/cm^3?

2. The density of nickel is about 9 g/cm^3. What is the volume of a block of nickel with mass 63 g?

3. The density of mercury is 13.6 g/cm^3. What is the mass of 200 cm^3?

4. The density of air is 0.0012 g/cm^3 or 1.2 kg/m^3. What is the mass of the air in a room 8 m long, 5 m wide and 3 m high?

5. Ten measurements are made of the density of gold. They are, in g/cm^3,

19.34	19.30	19.30	19.31	19.40
19.30	19.32	19.30	19.33	19.32

What is the average value?

6. A block of material A measures $5 \text{ cm} \times 5 \text{ cm} \times 8 \text{ cm}$ and it has a mass of 500 g. A block of material B measures $2 \text{ cm} \times 2 \text{ cm} \times 5 \text{ cm}$ and it has a mass of 170 g. Which is the denser material, A or B?

7. Measurements on various metals reveal that the mass of 1 cm³ is as follows:

brass	8.3 g
copper	8.9 g
iron	7.8 g
aluminium	2.7 g

Eight boys experimented with a block of metal which was 1 cm × 1 cm × 2 cm. Each measured its mass and got the following results:
16.2, 16.4, 16.6, 17.0, 26.6, 16.4, 16.6, 17.0 g.
What do you think is the metal of which the block is most likely to be made? Give your reason.

8. In an experiment to measure the density of air, it was found that the large cubical plastic container (measuring 30 cm × 30 cm × 30 cm) originally had a mass of 460 g. When air had been pumped in under pressure, the container had a mass of 466 g. When the 'extra air' was let out and collected as shown below, it was found that there was enough to fill a cubical Perspex box, 10 cm × 10 cm × 10 cm, five times.

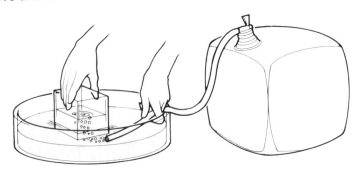

a. What is the mass of the 'extra air'?
b. What is the total volume of the 'extra air' released?
c. Calculate the mass of 1 m³ of air at normal pressure.
d. Someone thought it would be better to squash the container flat to see how much extra air could be driven out in this way. How many extra boxfuls of air would you expect to be able to collect?

Chapter 6 **Further measurements**

Scientists are often concerned with measurement and we have already learned something about measuring lengths, areas, volumes and masses. In the last chapter we learnt how to measure densities. In this chapter we shall think about some more ways of making measurements. Sometimes an experiment will not be very successful at first; then it is our task as scientists to think of ways of improving it. If possible, you should try all the experiments yourself; you can always compare your results afterwards with those of other people — scientists like doing that as well.

Experiment 6.1 Finding the thickness of a penny

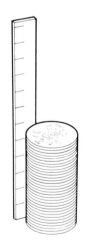

First make a guess what is the thickness of a penny. Then try to measure it using a plastic measuring rule marked only in centimetres.

Neither of the above are very precise. It would be better to use a rule with a millimetre scale on it. How precise can you be now? How can you improve the experiment further? The drawing gives a clue.

What is the height of ten pennies? What does that tell you about the thickness of one penny?

See if you can measure the height of 100 pennies. What does that measurement tell you about ten pennies? About one penny?

Would there be any point in going further?

A project to try at home

You might like to investigate how the thickness of 5 p pieces and the old shillings varies with their age. Collect (or borrow) as many as possible and measure the thickness as accurately as you can to see how it changes.

Experiment 6.2 Finding the thickness of a piece of paper

Measure the thickness of a piece of paper as accurately as you can.

The thickness of paper in books and magazines varies considerably. Investigate how they differ. Find the thickness of a page in a Bible and compare it with the thickness of a page in a glossy magazine.

Measuring mass

Experiment 6.3 A simple balance to find the mass of a parcel

The object of this experiment is to find the mass of a small parcel using a simple balance. Set up the balance as shown below.

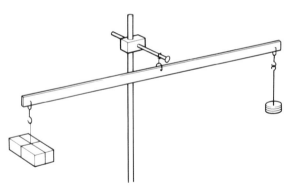

Put the parcel on one side of the balance and the hanger on the other side. Add 100 g masses to the hanger and try to work out the mass of the parcel.

It will probably be very difficult to get it to balance exactly. Does this matter? Can you still estimate the mass of the parcel?

Experiment 6.4 Using the simple balance to find the mass of a letter

Use the same balance to find the mass of a letter. It is much lighter than the parcel so you will need to use a hanger with 10 g masses instead of the 100 g masses.

It may be very difficult to get it to balance. If so, use a bulldog clip instead of the central hook. Can you estimate the mass?

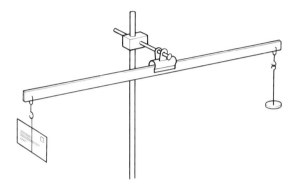

Suppose you need to know if the mass is greater or less than 25 g, but your masses on the hanger only give 10 g, 20 g, 30 g and so on. Can you think of a way of doing this?

Perhaps the drawing below will give you a hint of how it might be done.

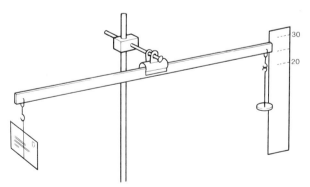

The above apparatus is quite good for finding the mass of an envelope, but it would not find the mass of a dead fly or a hair. It looks as though we need something different for that.

Experiment 6.5 Making a microbalance

You will need the equipment shown below.

Put the screw in the end of the drinking straw and fit the other pieces together as shown. Before sticking the needle through the straw the approximate position can be found by

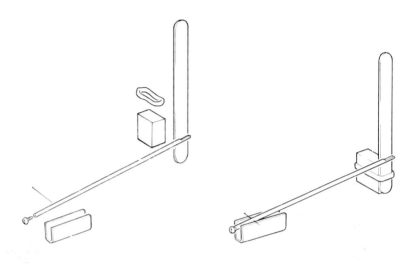

centre   insert needle here

balancing the straw on the needle. The needle should not go through the centre of the straw, but above it (see drawing). The screw can be used to adjust the balance. Squash the other end of the straw to make a little platform on which to put things to be measured.

Do not worry if you damage your straw: you can always have another one.

When it is assembled and balancing, put a dead fly or a hair on it, and see if it is working. You may find that draughts are a nuisance: if they are, make a shield with a screen or a large book.

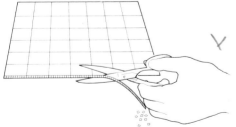

Experiment 6.6 Using the microbalance

To do any measuring with your microbalance, you will need some masses. You have some graph paper and some scissors with your apparatus. You can cut up some small squares from the graph paper and these will make very useful masses.

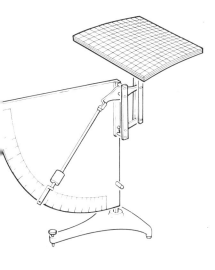

Mark the position of the balance when there is no mass on it. Put on one small square and mark the position, put on a second square and mark that position, and so on. Go on putting on more masses and mark all along the scale. If you put the hair on the balance instead of the squares, you will be able to read off the mass of the hair, measured in 'small squares'.

How can you find the mass of the hair in grams?

You need to know the mass of each 'small square'. The easiest way to find this is to put 100 sheets of the graph paper on a balance and find their mass. It is then easy to find the mass of 1 sheet. Count up the number of large squares on the sheet and you can calculate the mass of one of them. But there are 100 'small squares' in one large one. So you can find the mass of one small square. It is usually convenient to measure it in milligrams (1 mg $= \frac{1}{1000}$ g).

Can you now find the mass of the hair in milligrams?

Sensitivity of a microbalance

We call a balance very sensitive if a small mass added to it moves the pointer a large distance. The sensitivity of a microbalance depends to some extent on where the needle is put through the drinking straw. The sensitivity is low if the needle is put near the edge of the drinking straw, it gets higher the closer the needle gets to the centre of the straw. If you have time, try changing the sensitivity of your microbalance.

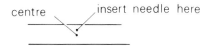

Experiment 6.7 Measuring the thickness of aluminium leaf

This is a difficult experiment and will require some thought. The aluminium leaf is very thin and it is not possible to pile it up and measure the thickness of the pile as was done for sheets of paper.

Start with a square piece of leaf, 5 cm × 5 cm. We know its length and breadth, but want to know its thickness.

We cannot measure its thickness directly. Is there anything else we can measure? Its mass. This is very small, but it can be found with the microbalance.

If we know its mass, can we find its volume? Yes, because we have already found the density of aluminium in an earlier experiment. As we already know the length and breadth of the original piece of leaf, we can now use the volume to find the thickness.

This experiment can be represented as follows:

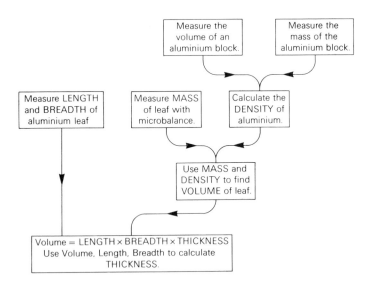

This is an example of a *programmed* piece of scientific work. It will be an achievement if you succeed in getting an answer.

Measuring time intervals

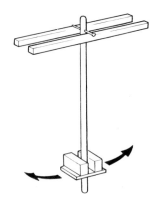

Time is another quantity that scientists need to measure. The usual unit is the *second* (abbreviated s) though there will, of course, be occasions when it is necessary to measure in minutes, hours, days or years.

The time between two events can be measured with a clock, preferably one with a second hand. Stop-watches are particularly useful for timing, but they are expensive. You can do some good timing using a broomstick pendulum as shown on the left. It shows what good physics can sometimes be done with very simple apparatus.

Experiment 6.8 Investigation of a broomstick pendulum

How can you measure the time between swings of a broomstick pendulum? It is not at all easy to measure with any accuracy the time between two clicks with a watch, or even with a stop-watch. The best method is to find the time that lapses between the beginning and the end of ten intervals of time, and then to divide the answer by ten.

Questions for class discussion on the broomstick pendulum

1. Why is it more accurate to time ten intervals and divide the answer by ten than it is to time one interval?

2. When counting ten intervals, why must you be careful not to count the first click at which you start timing, as one, the second as two and so on up to ten? Why should you count the first click at which you start timing as zero, the next as one and so on up to ten? (In practice, it helps sometimes to have a count-down to zero.)

3. Does it make any difference to the time between clicks if the swings are very small ones instead of large ones? It is not likely that you can answer this by just talking about it. It will be necessary to do an experiment to decide.

4. Why does your last answer mean that, even though the swings are getting less and less, the time interval between clicks does not change?

Timing things

You should use a stop-watch or your broomstick pendulum to time different things: the time between hand-claps, the time for a book to fall, the time it takes to measure the mass of an aluminium block, the time to walk round the room or to run round the building.

Try estimating time intervals. You can become quite good at counting seconds with a little practice. One way is to say 'apples and pears one', 'apples and pears two' and so on at a normal speaking speed. (Some people have their own words, for example 'Mississippi one', 'Mississippi two',

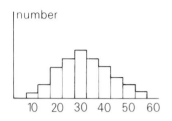

number

10 20 30 40 50 60

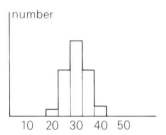

number

10 20 30 40 50

etc.) It does not matter what you use, but you should practise with it and you will soon find the right speed at which to say the words so that you can count quite accurately to 30 seconds.

Experiment 6.9 Class experiment estimating 30 seconds

An interesting game can be played in which everyone in the class estimates 30 seconds. The teacher has a watch and starts everyone together. As soon as a pupil thinks 30 seconds have gone by he raises his hand quietly. (It is better to do it with your eyes shut and to raise your hand so that others do not hear.)

The first time it is likely that someone will put his hand up after 20 seconds or even earlier, and some may not raise theirs for 50 seconds. There might be a spread like the first graph.

After you have played the game several times, everyone will be much better at estimating. Perhaps the spread will be more like the second graph.

Can you see the difference?

Statistics

A statistics frame is a useful way of showing a spread of results. The following experiment will illustrate its use.

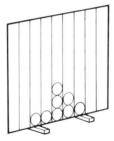

Statistics frame with discs

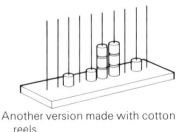

Another version made with cotton reels

Experiment 6.10 Class experiment with statistics frame

Everyone in the class should find his mass in kilograms. (Some bathroom scales are useful for this.) Every pupil should be given a disc to place in the statistics frame. Suppose the first slot in the frame represents masses up to

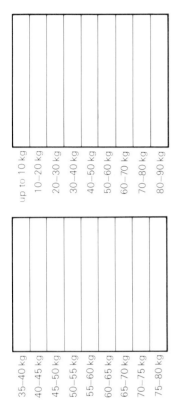

10 kg. The next slot masses between 10 kg and 20 kg. The next between 20 kg and 30 kg. And so on up to 90 kg.

Each pupil then puts his disc in the slot appropriate to his mass. You will see that we get a spread. But you will notice we soon get one or two slots over full. What can we do about it?

Let us try a different spread. Suppose the first slot is 35–40 kg, the next 40–45 kg, the next 45–50 kg, and so on. This spreads out the commonest masses.

Other statistical experiments

1. Let everyone in the class measure his height (or the length of his foot, or the size of his waist). Put the results on the statistics frame.

2. Let everyone in the class measure the mass of the same block of aluminium. Not everyone will get the same answer: there will be a spread of readings. Show these on the statistics frame.

3. If your school has a lot of 500 g masses (or 1 kg or 200 g masses), measure the mass of each of them on a lever arm balance. It is unlikely that all the readings will be the same. Show the spread on the statistics frame.

Averages

Experiments 2 and 3 above show that we must be very careful about using phrases like 'absolutely accurate' when making measurements: there is usually a statistical spread. The best you can do is to get an average value.

The usual method for finding an average is to add all the measurements up and divide by the number of measurements. Thus the average of 501 g, 507 g, 502 g, 502 g, 503 g is obtained as follows:

$$\text{average} = \frac{501 + 507 + 502 + 502 + 503}{5}\,\text{g}$$
$$= \frac{2\,515}{5}\,\text{g} = 503\,\text{g}$$

The average value is therefore 503 g.

There is in fact an easier way to work out the average of those five numbers. Can you think what it is?

Estimating games

Being able to estimate distances and times is very helpful in scientific work. You can get practice in this when out for a walk. Before you go, use a metre rule to measure the length of your pace. What distance in metres will be the same as ten paces? What distance will be the same as 100 paces?

1. Choose something ahead of you on your walk, like a lamp-post or a tree, and estimate how many paces you think it will be until you reach it. Then pace it out and see who has the best estimate.
2. Then start to make estimates in metres. Try it with objects which are quite near and ones which are much further away.
3. Another game is to estimate the time it will take to walk to the tree or the lamp-post, if you walk at a steady speed.
4. A more difficult game is to work out the speed of a car on a road. You must first pace out 100 metres along the road, and then estimate the time a car takes to travel that distance. From this you can work out its speed.
5. Along the side of a railway track there are usually quarter-mile posts (see photograph) and $\frac{1}{4}$ mile is very close to 400 m. If you look out of a train and estimate the time taken between two of these posts, you can find the speed of the train.

Quarter mile post

Time between $\frac{1}{4}$ mile posts in seconds	Time to travel 1 mile in seconds	Speed miles per hour	Speed kilometres per hour
10	40	90	144
15	60	60	96
20	80	45	72
30	120	30	48
60	240	15	24

Measuring temperatures

Before we leave measurements, we ought to mention one more, temperature measurement. You will be studying heat later in your course: this is merely a first look at one aspect of it.

In scientific work we measure temperature on a scale called the Celsius scale. On this the temperature of melting ice is 0 °C and the temperature of boiling water is 100 °C. If the temperature outside is below 0 °C, water will be freezing and turning to ice. 5 °C would be a very cold day outside, whereas 30 °C would be very hot indeed.

You are probably quite good at telling whether a bath is hot or not by putting your toe in it. But actually our body is not very good at estimating temperatures. Try the following experiment.

Experiment 6.11   Measuring temperature

Take three bowls of water, one nearly full of very cold water, one containing tepid water and the third full of hot water. Put one hand in the cold water, one hand in the hot water, and keep them there for a minute. Then put both hands in the tepid water. How does the tepid water feel to each hand? One hand tells you it is warm, the other tells you it is cold!

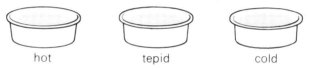

hot   tepid   cold

To measure temperature we use a thermometer. Put a thermometer in the cold water and another in the hot water and see what they read. Then put them both in the tepid water and you see that they read the same. This is one reason why scientists prefer thermometers to hands when measuring temperature.

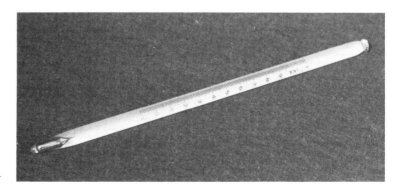

Thermometer

Chapter 7 **Forces**

We are always pushing and pulling. We have a special name for a push or a pull : we call it a *force*. Forces are everywhere around us and a few examples are shown in the drawings opposite. Look at each picture and think about what forces are acting.

How do we recognise a force? If you put a football in the middle of a field, it will remain there without moving until a force is applied to it. The force may be caused by the wind blowing it, by someone kicking it or by someone picking it up. In each case a force causes motion.

A force can also decrease motion or stop it. A moving football can be stopped by the force exerted when it hits a player. A parachute attached to a jet plane when it is landing causes a force to be exerted on it and this slows it down. We can recognise a force as something which changes motion.

If a football hits a wall, it will bounce off it. If it hits the wall at an angle, it will bounce off at an angle; the force of the wall has in this case deflected the motion. The direction of the force is clearly important. When you kick a stationary ball, it moves in the direction of the force. When it is picked up, it moves in the direction of the upward force applied.

Forces in equilibrium

We can recognise forces by their ability to start motion, or to change motion. But if you sit on a chair there are forces acting on you and yet you do not move. In this case the forces just balance and produce no motion. The forces are said to be *in equilibrium*. Sitting on the chair, there is a downward force on you due to gravity (your weight) and this is exactly balanced by the upward force exerted on you by the chair.

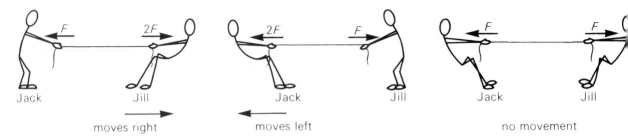

Jack   Jill   Jack   Jill   Jack   Jill

moves right   moves left   no movement

The drawings above show two people pulling on a rope. In the first, Jill pulls with twice the force of Jack: she will pull Jack to the right. In the second, Jack exerts the bigger force so the motion is to the left. But in the third drawing the forces are equal, they are in equilibrium and there is no motion.

If a body is in motion and there is a frictional force acting on it, it can be brought to rest by that force. If there is no force on it, the moving body will go on for ever. It may be surprising to learn that a body could go on for ever without any force acting on it, but think of a block of wood moving over the ground. Frictional forces soon bring it to rest. But put it on smooth ice where the frictional force is much smaller, it will travel much further before the friction stops it. Now imagine the frictional force getting smaller and smaller, the block will travel further and further. Imagine then what happens when there is no frictional force: the block would go on for ever.

A moving body will have its motion changed by forces acting on it, but if the forces on it are in equilibrium then it will continue to travel at a steady speed as if no forces were acting.

For example, if the forward thrust due to the engines of a plane is equal to the backward force due to air resistance, the forces are in equilibrium and the plane moves at a constant speed. If they were not in equilibrium, the plane would speed up or slow down depending on which was greater.

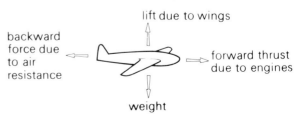

lift due to wings

backward force due to air resistance

forward thrust due to engines

weight

Kinds of forces

The most obvious forces of which you are aware are pushes or pulls exerted through the actions of our muscles, but there are many other different kinds of force.

Magnetic forces

Magnetic forces can be felt by holding the ends of two magnets together. Sometimes the forces are attractive ones, sometimes they are repulsive forces. To find out about these it is best to do your own investigation as in the next experiment.

Experiment 7.1   Investigation of magnets

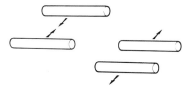

Take two cylindrical magnets and first arrange them on the bench so that they attract each other. What happens if you turn one of the magnets round? What happens if you then turn the other round as well?

Hold the magnets, one in each hand. Bring two of the ends together. What happens? What happens if you bring the other two ends together? Turn one of them round. What happens when the ends are brought together now?

It is usual to call one end of a magnet a North-seeking pole and the other a South-seeking pole. Why do you think this is? Can you suggest an experiment you can try to show why one end is called North-seeking?

The rule for magnets is that like poles repel, unlike poles attract. Can you see what this means? Did your experiments agree with it?

Horse-shoe magnets also have poles. Can you think of a way to show that one end is a North-seeking pole and the other a South-seeking pole? (*Hint:*   You already know that one end of your small cylindrical magnet is a North-seeking pole and 'like poles repel'.)

A good demonstration to show magnetic forces is given by mounting two horse-shoe magnets on trucks as shown below. Push the trucks together and see what happens. What would happen if one of the magnets was put the other way up? What would happen if they were both turned the other way up?

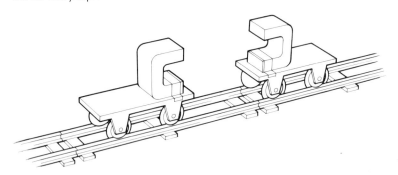

Electric forces

Later in your course you will learn about electric forces. You will find there are two kinds of electric charge which for convenience we call *positive* and *negative*. There is a repulsive force between two positive charges, there is also a repulsive force between two negative charges, but an attractive force between a positive and negative charge. We must leave detailed discussion of this until later, though it would be interesting to do one experiment to show it.

Experiment 7.2   Electric forces shown with balloons

Blow up two rubber balloons. Attach a long thread to each and hang them up from a horizontal thread as shown above.

Rub one of the balloons with your sleeve. Then rub the other balloon similarly. What happens when the balloons come near to each other? What happens when you put your sleeve near either of the balloons?

This is only a brief look at electrical forces. We will return to them later.

Gravitational forces

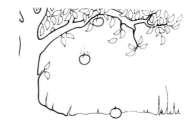

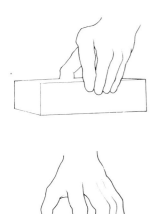

We are all aware of the gravitational force pulling everything towards the centre of the Earth. It is this gravitational force which causes the apple to fall from the tree or a brick to fall to the floor when we let go of it. It is this gravitational force between a body and the Earth which we call the *weight* of the body.

At a very young age, a small child realises that to hold up a ball — to keep it in equilibrium — he must exert an upward force on it to balance the weight.

It was Sir Isaac Newton who first realised that there is a force of attraction between any two masses. It is usually a very small force, but if one of the masses is the Earth then the force is very much greater and we are all aware of it.

Since the mass of the Moon is much less than that of the Earth, the gravitational force on objects is less on the Moon. This means that the weight of a given mass will be less on the Moon than it is on the Earth: the weight of a kilogram of lead on the Moon is about one-sixth of its weight on the Earth.

Impact forces

A stationary body can be set in motion if a moving body collides with it. The moving body therefore exerts a force on it, and such a force is called an *impact force*. A moving hammer can knock a nail into a wall, a stream of water can wash away its banks, the jet of water from a hose pipe can move soil or a pile of sand, a strong wind can blow trees over. Thus solids, liquids and gases can all exert impact forces.

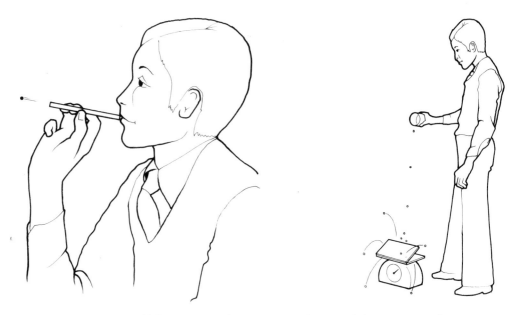

When a pea from a pea-shooter hits a target, it exerts a small impact force on it. This lasts for a very short time, but if there is a continuous stream of peas it will seem like a continuous force. This is how a stream of water or a gust of wind exerts a force. Each molecule exerts an impact force. As there are many such molecules all the impact forces will add up to give a large force.

Strain forces

When you pull a piece of rubber (for example, an elastic band) by applying a force with both hands, there is a small movement and the rubber becomes longer until equilibrium is reached. You become aware of a force or *tension* in the rubber. If you pull a bit harder, it stretches a bit more and the tension becomes greater. The more you pull, the more it stretches and the greater the tension in it, provided of course you do not pull too much and break it. If you let go, there is no longer equilibrium, the tension force sets the parts of the rubber in motion again so that it returns to its original length.

Such strain forces are felt not only by stretching. If you take a piece of rubber and twist it, you will feel similar forces. When you let go, the rubber goes back to its original shape.

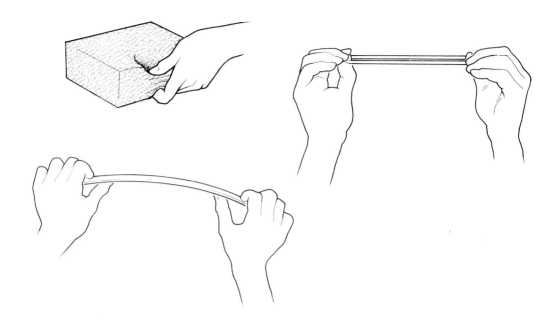

Strain forces are also experienced when compression takes place. This can be shown by mounting two springs on the railway trucks as illustrated below. What happens when they are pushed together?

If you take a strip of metal, like a hacksaw blade, or a long piece of wood, like a metre rule, and bend it, you will feel the strain forces. The more you bend it, the greater the force.

Bodies which return to their original shape when the force is removed are called *elastic* bodies.

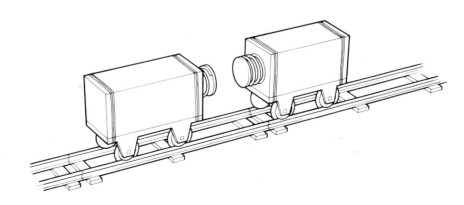

Golf ball squashed on impact

It is these strain forces which explain why a ball bounces. A moving ball hits the floor, there is an impact force which deforms the ball, flattening it slightly. The impact force ceases and the strain force restores the ball to its original shape causing the ball to move away from the floor.

Frictional forces

Some of the most common forces are *frictional forces*. If you put a block of wood on a table and give it a small push, it does not start to move because there is a frictional force in the opposite direction to the push. If you push a bit harder the frictional force also gets bigger and again stops the motion. The moment you stop pushing, the frictional force immediately ceases as well. There is however an upper limit to the friction. When you push much harder, you pass this upper limit to the friction and motion begins.

Friction therefore has some strange properties. It cannot start a body moving, it can only oppose motion and the size of the force can vary up to a certain maximum.

We usually think of friction as a nuisance: we oil machinery to reduce it. But most of the time it is essential to us. Think how difficult it is walking on a slippery surface: without friction it would be impossible to walk on the floor.

Questions for class discussion

1.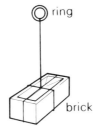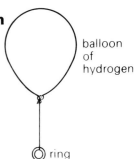

Suppose you put your finger through the ring in each of the above. What force will you feel? In what direction will it act? Will there be any change in the force if you move your finger very slowly upwards? Or downwards?

2.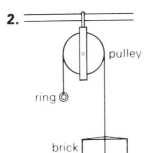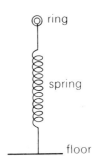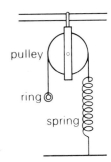

If you put your finger through the ring in each of the above, what force will you feel? In what direction will it act this time? Does it make any difference to the force if you move your finger slowly upwards? Or downwards? What is the effect of the pulley?

3. Give examples of each of the following:
a. a force exerted on something which starts it moving,
b. a force exerted on something which stops it moving,
c. a force which changes the direction of something moving,
d. a frictional force exerted on a moving body,
e. a frictional force exerted by a liquid,
f. a frictional force exerted by a gas,
g. a strain force when something is stretched,
h. a strain force when something is compressed,
i. a gravitational force on a solid body at rest,
j. a gravitational force on a moving body,

71

4. When a skater is moving on ice, there is very little friction. How is it possible for him to turn a corner?

5. When an aeroplane is flying horizontally at a steady speed, the forces on it are in equilibrium. The aeroplane has a downward force on it, its weight. What does this tell us about the upward force on the aeroplane?

6. Look at the drawings on page 62 and describe what forces are acting in each picture.

Forces occur in pairs

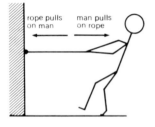

It is important for an understanding of forces to realise that they always occur in pairs. If Jack pulls on a rope, the rope also pulls on Jack with the same force. If a block of wood lies on a table, the block exerts a downward force on the table; but the table also exerts an upward force on the block of wood. When you are walking on a floor, your foot exerts a force backwards on the floor and the floor exerts a forward force on you.

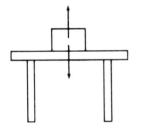

The Earth pulls on the Moon with a gravitational force: it is this force which keeps the Moon in its orbit. But the Moon pulls on the Earth with an equal and opposite force: it is this force which is responsible for the tides.

Questions for class discussion

1. The Earth pulls on an apple and this force causes the apple to fall to the ground. Forces occur in pairs. The same force must therefore act on the Earth due to the apple. Why does not the Earth appear to move towards the apple instead of the other way round?

2. Explain the following.
a. Two equal masses are joined together by a string which is hung over two smooth pulleys.
The string suddenly breaks at A and both masses fall in the same direction to the ground.

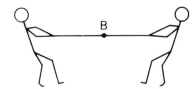

b. Two boys are pulling each other with a rope. Suddenly the rope breaks at B and they both fall over backwards in opposite directions.

Measuring forces

We have already seen that scientists like to measure those things with which they have to deal. Forces are no exception, and they are measured in *newtons*. You should experience for yourself what a force of 1 newton is like by pulling on the forces demonstration box. (The abbreviation for 1 newton is 1 N.)

Newton spring balance

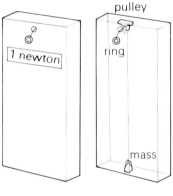

(The second drawing shows how the box is made. The string passes over a pulley and on the lower end is a mass of about 102 grams. The downward gravitational force on this mass is almost exactly 1 newton. So pulling on the ring on the front of the box gives the feel of 1 newton.)

Forces can often be conveniently measured using a newton spring balance. This has a scale on the side marked in newtons and the position of the pointer tells you the force applied.

Experiment 7.3 Using a spring balance

Use a spring balance, marked in newtons, to measure a lot of different forces.

Find out what size of force is necessary for the following:

a. to pull a block of wood along a table,
b. to pull a table or chair across the room,
c. to pull a drawer open,
d. to pull a door open.

Use the newton spring balance to measure the gravitational force on a shoe, on a book, on other objects in your laboratory.

What happens if you join two spring balances together? Fix one of them so that it hangs vertically. Attach the other spring balance to it and pull with a force of 5 newtons. What is the reading on each spring balance?

By hanging a 1 kg mass on the spring balance, measure the gravitational force on it. Do the same with a $\frac{1}{2}$ kg mass if one is available.

The strength of the gravitational field

We say there is a *gravitational field* anywhere a gravitational force is felt. We are in a gravitational field everywhere on the Earth's surface.

The previous experiment will have shown that when you hang up a mass of 1 kg the downward force on it due to gravity is approximately 10 newtons.

Similarly, the downward force on 2 kg is 20 newtons, on 3 kg 30 newtons and so on. We can say that the strength of the gravitational field is 10 newtons on every kilogram, or 10 newtons per kilogram (10 N/kg).

On the Moon, the gravitational field of the Moon is about one-sixth of this. In other words, the force on 1 kg would be only 1.7 newtons. This downward force is what we call the *weight* of the body, so the weight of an object on the Moon would be less than that of the same object on the Earth. As weight is a force, it should always be measured in newtons. Unfortunately, suppliers of groceries do not always use the correct scientific terms: a packet of sugar may be labelled

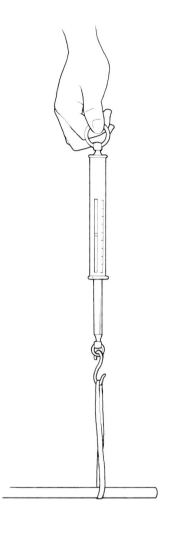

'weight 250 g', or 'weight 1 lb', whereas it should read 'mass 250 g', or 'mass 1 lb'.

You now know that a mass of 1 kilogram will have a weight of approximately 10 newtons on the Earth and its weight on the Moon is approximately 1.7 newtons. The mass is always the same, the weight will be different in different gravitational fields.

Experiment 7.4   Stretching elastic bands

Hang up a long elastic band using a rigid support. Measure its length.

Connect a newton spring balance to the elastic band. Pull on the balance so that a force of 1 newton is applied to the elastic band. Measure its new length. Increase the force to 2 newtons, measure the length again. Continue to take readings with greater forces.

If you release the force, does the elastic band go back to its original length?

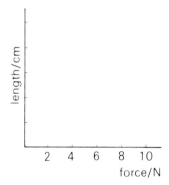

It is usually easier to see what is happening if you draw a graph. Put the force along the x-axis and the length along the y-axis. Plot each of your readings. What do you notice about the graph you obtain?

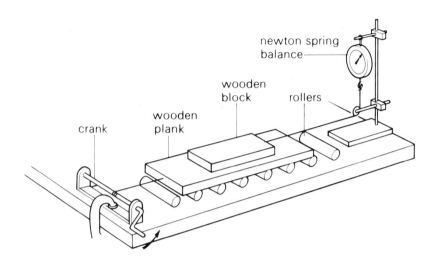

newton spring balance

wooden block

rollers

wooden plank

crank

Experiment 7.5 Investigating friction

The frictional force between a wooden block and a wooden plank can be demonstrated with the apparatus in the drawing. If the crank is turned, the wooden plank can be pulled steadily over the rollers. The spring balance will measure the frictional force on the block.

Measure the frictional force when the plank is pulled at a steady speed. Then put a second block on top of the first and measure the frictional force again. It will be greater. Add more blocks and each time measure the force.

You can also investigate the frictional force when you press down on the block with your hand. What happens if you turn the block on its side so that the area of contact is less?

Homework assignments

1. Read the topic book *Forces*, pages 13–17, and prepare a short talk on universal gravitation.

2. Read *Forces*, pages 18–21, and prepare a short talk on either sailing or flight.

3. Read *Forces*, pages 25–28, and write an essay on friction, including in it something about the possible cause of friction.

4. Read *Forces*, pages 36–37, and prepare a short talk on the muscular forces in your arm.

Chapter 8 **Investigation of Springs**

In this section you will be doing a lot of investigation, finding out how springs behave. At the end you should know quite a lot about them and be able to predict how they will behave in any future experiments you do. In order to help sort out what you find, keep a careful record of your investigations in your notebook. Good scientists always do that.

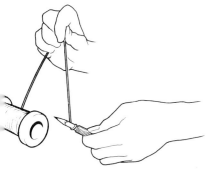

Experiment 8.1 Home-made springs of copper wire

To begin this investigation of springs, start by making your own spring with copper wire. You will need about 80–90 cm of wire which should be wound into a spiral on a pencil. The spring should be made by turning the pencil round, not by twisting the wire. Make a twisted loop at both ends so that you can hang up your home-made spring using a nail as shown in the drawing on the left.

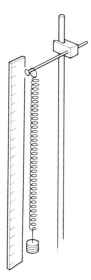

a. Use a metre rule (or a half-metre rule, whichever is more convenient) to measure the length of your spring before anything is attached to it.
b. Attach a small hanger to the lower end of the spring. What happens? Is there anything you can now measure? What is the best way to hold the rule? What is the most convenient way to measure the length?
c. Add a 10 g mass to the hanger. What is the total load now? Is it 10 g or more? (*Hint*   What is the mass of the hanger?) What is now the length of the spring?
d. Take off the 10 g mass. What does the length go back to? What does it go back to if you take off the hanger as well?
e. Put the hanger back on again. Measure the length once more. Add 10 g and re-measure. Add more masses, keeping a record of the lengths.
f. After you have loaded up with several masses, find out what

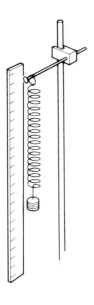

happens to the original length when you take the hanger off.

g. Go on adding more masses. Is there any limit to the number you can add? Describe what happens.

h. If time allows, make another spring, winding it on the same pencil, but using twice the length of copper wire and making the spring twice as long. Load up this spring in the same way. What difference does this new length make?

Experiment 8.2   Investigation of steel springs

Carry out a similar investigation using the steel springs provided instead of your home-made copper spring.

This time you should use larger hangers, which have a mass of 100 g. Add masses which are each 100 g. You may find it makes it easier if you fix the rule in a clamp, but that is for you to decide.

Do not worry if the springs get damaged, your teacher will always supply you with another.

When you have finished the investigation, write down a description of how the behaviour of the steel spring differed from that of the copper spring.

Plotting results

It is often much easier to see what is happening at a glance if you plot the results on a graph instead of just making a list of measurements. You can plot the length of the spring against the load, as shown in the first graph below.

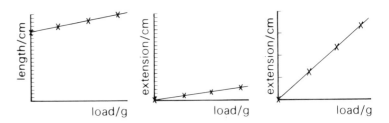

Sometimes it is more convenient to plot the extension against the load instead of the actual length. By *extension* we mean the amount the length has increased or extended.

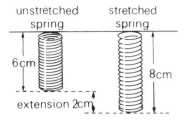

unstretched spring · stretched spring

6 cm

8 cm

extension 2 cm

If a spring has an unstretched length of 6 cm and a load attached to it stretches it so that the new length is 8 cm, then the extension is (8−6) cm or 2 cm. If you plot extension against load, you will get the second graph shown above.

An advantage of plotting extension against load is that you can now change the vertical scale and get a graph like the third one shown above.

A simple rule for stretching springs

Use the readings you obtained for your steel spring in Experiment 8.2 to plot a graph of extension against the load applied. Can you see any simple rule about the way the spring stretches? (*Hint:* What happens to the extension when you double the load, what happens if you treble the load?)

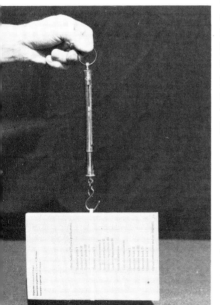

Newton spring balance

Further experiment

When you applied a small force to a steel spring (by hanging a small load on it) and you removed the force, you found that the spring went back to its original length. But will this happen however large the load? Investigate what happens when you apply much larger loads. What happens to your plot of extension against load when these bigger loads are applied? You will certainly damage the spring by having to test it like this, but scientists have to do this if they want to know all about how a spring behaves.

The newton spring balance

In the last chapter you learnt to use a spring balance to measure forces. Your experiments in this chapter will have shown you how they work. The balance contains a spring which stretches when a force is applied. The rule for the stretching of springs tells that the extension increases steadily as the force is steadily increased. You can therefore put a scale on the side which tells you directly what the force

first spring

second spring

is. Most spring balances have a stop which prevents the spring inside being overpulled. We usually refer to the stage at which the spring does not return to its original length as the *elastic limit* of the spring. The stop prevents us reaching the elastic limit of the spring inside the spring balance.

Experiment 8.3   Investigation of springs in series

Take two similar springs and measure how much each stretches when a 100 g load is attached to them. (If the springs are similar, the extension should obviously be the same in each case.)

Then join the two springs together as shown in the drawing. (We call this joining them *in series*.) How much does each spring stretch when a 100 g load is attached to the bottom end? How does this compare with the extension when a 100 g load was added to each spring separately?

On the same sheet of graph paper, plot the graph of extension against load, first for one spring on its own and then for the two springs in series.

Make a guess what the graph might look like if there were three springs in series. Then do an experiment to see if your guess was correct.

Experiment 8.4   Investigation of springs in parallel

Hang the two springs side by side as shown on the left. (We call this joining them *in parallel.*) Investigate how the extension varies with the load in this case.

Plot the graph of extension against load on the same graph paper as a similar plot for one of the springs on its own. How does this differ from the result obtained in Experiment 8.3? Why does it differ from that result?

What do you think would happen if the load were not attached at the mid-point, but closer to one of the springs? Try it and see if it agrees with your answer.

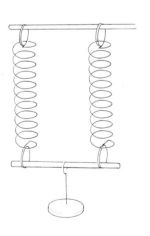

Stretching copper wire

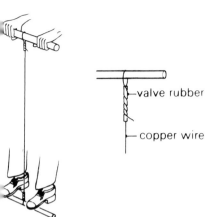

valve rubber

copper wire

In this chapter we have so far only considered springs, but most things stretch a bit when force is applied. You can feel this for yourself if you take some very thin copper wire, just over a metre will do. If you try to pull it as it is, it will probably pull through your fingers. A good way to hold it is to wrap each end round a pencil and twist the ends round a piece of valve-rubber as shown. Hold one pencil between your feet and the other in your hands. Pull on the wire and see if you can feel it stretching. You can probably feel the 'cheesy' effect as it stretches.

If you break your wire, look at the broken ends with a hand lens. What do you see?

Experiment 8.5   Stretching copper wire

We have already mentioned that scientists like to make measurements and it would be interesting to know how much copper does stretch. Different loads can be put on the long length of wire shown in the apparatus below. To show how much the wire stretches, a thread is attached to the wire and given one turn round the knitting needle as shown. A weight hanger with about 40 g on it keeps this thread taut and the point of the knitting needle is pushed through a drinking straw.

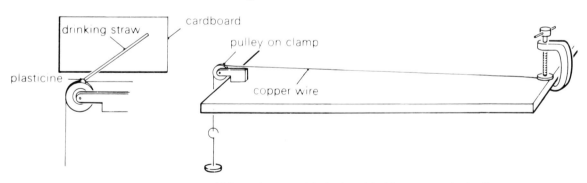

drinking straw

cardboard

pulley on clamp

plasticine

copper wire

What do you think the drinking straw is for?

See what you can find out about the stretching of copper wire when different loads are added. Does it go back to the original length when the load is removed? Does a long piece of wire stretch more than a short piece? What is the difference if your wire is not quite so thin? When it breaks, does it always need the same load to break it? Does it always break at the same place? What does the break look like?

If you have time, try some other substances: nylon thread, cotton, thin string, other kinds of wire.

Questions for homework or class discussion

1. A spring is 4 cm long. When a load of 200 g is attached to it, the new length is 6 cm and it returns to 4 cm when the load is removed. What will the length be when a load of 100 g is attached instead of the 200 g load? What will it be for a load of 50 g?

2. The extension of a 6 cm spring is 4 cm for a load of 400 g. The spring returns to 6 cm when that load is removed. What is the extension for a load of (a) 100 g, (b) 200 g, (c) 600 g, (d) 60 kg?

3. A spring has an extension of 4 cm for a load of 800 g. What will be the extension when a second similar spring is connected in series with it and a load of 800 g is attached at the bottom?

4. What would be the extension if the two springs in the previous question were connected in parallel and the 800 g load were attached to the middle of a light rod connecting their lower ends?

5. Spring X is similar to spring Y except that X is twice as long as Y. If a load of 1 kg stretches Y by 10 cm, by how much will the same load stretch spring X?

6. Spring A is 10 cm long when no load is attached, but it stretches by 8 cm when 400 g is attached. It returns to its original length, and an unknown load is then attached to it (instead of the 400 g). The new length of the spring is 13 cm. What was the unknown load?

7. A spring stretches 12 cm when a load of 600 g is attached to it on the Earth. The gravitational field on the Moon is one-sixth of its value on the Earth. How much would the spring stretch if 600 g were attached to the same spring on the Moon?

8.

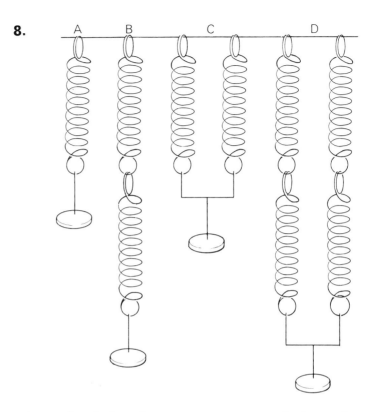

In each of the above arrangements, all the springs are similar. In each case the load is exactly the same. Write down which of the four arrangements will show the greatest extension? Which will show the least extension? Say what you can about the extension of the other two.

9. A mass of 1 kg is attached to the end of a spring so that the spring stretches 2 cm. The weight of the mass pulling downwards on the spring is a force of 10 newtons and it is this force which causes the spring to stretch.
a. How big is the force exerted by the spring on the ceiling?
b. What is the direction of the force which the spring exerts on the ceiling?
c. If the experiment were done on the Moon, would you expect the spring to stretch more, less or the same amount? Give the reason for your answer.

ceiling

1kg

10. A boy fixes the top of a spring to a hook. To the other end he attached various masses. Every time he changes the masses, he measures the length of the spring. The following is a record of his measurements, in the order in which he made them.

Load/g	Length/cm
0	10.0
200	12.4
400	14.8
600	17.2
400	14.8
0	10.0
600	17.2
800	20.4
1 000	26.0
400	19.0

a. What would have been the length if the third reading had been 300 g?

b. The first time there was a mass of 400 g, the length was 14.8 cm. It was the same the second time. Why was it different the third time?

c. If he had taken the mass off after the last reading so that the load was once again 0, would the length have been 10 cm? Give a reason for your answer.

Chapter 9 **The turning effect of forces**

It is very difficult, if not impossible, to loosen the nut on the hub of your bicycle wheel with your fingers. It is necessary to use a spanner. If it is very difficult to loosen, where do you hold the spanner? In the middle or at the end?

We have already discussed forces. We know that their size matters, it also matters in which direction they are applied, and we will see in this chapter that it also matters at what point they are applied.

If a man pushes on an open door with a large force near the hinges, the door can be held by quite a small force exerted by a boy pushing in the opposite direction at the edge of the door.

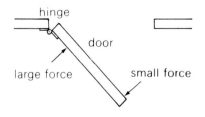

hinge

door

large force   small force

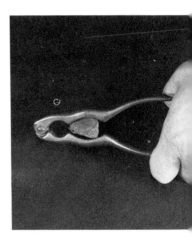

Coin and tin
Screwdriver and paint tin
Nutcrackers

The turning effect of a force depends very much on where the force is applied. A claw-hammer enables you to pull out nails you cannot move with your fingers; a screwdriver enables you to open a tin of paint; a coin helps you to open a tin of coffee; nut-crackers make it possible to crack nuts you could not break with your fingers; a crowbar (iron lever) lets you raise a crate you could not lift on your own.

Imagine a painter, standing on a plank of wood supported on two trestles while painting the ceiling. All is well when he stands on the plank between the trestles, but if he moves along the plank to the part which overlaps the end, what may happen? In this chapter we shall look further at the turning effect of forces.

Experiment 9.1 Investigation leading to the law of the lever

Set up the apparatus shown below. The wooden beam should be supported with the central groove resting on the top of the triangular support. (Such a support is sometimes called a *fulcrum*.) If the beam does not balance, put a lump of Plasticine underneath at some position so that it does balance.

Put some metal squares on one side and some on the

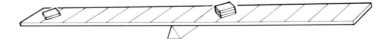

other so that the beam balances. You should put the loads so that they are one, two, three divisions out and not say, two divisions and a bit, because those fractions will make it more difficult to find out any law. It is probably best to put the squares diagonally on the beam rather than squarely.

Start by making the beam balance with two piles of squares, one on each side. You will not be able to get it to stay exactly balanced in mid-air, but the beam will tip over to one side or the other. It will be much like 'weighing sweets': find the point at which a very little more will tip the scale one way or the other.

Find out what you can about balancing the beam. Get four squares on one side balanced by two on the other; then get four squares balanced by one or by three. Keep a record of how many squares on the left at different distances are balanced by how many on the right, and at what distances. See if you can find some rule which tells you in advance whether two loads will balance. You may find that a table helps.

No. of squares	Distance	balanced by	No. of squares	Distance

Questions for class discussion

1. Two squares are placed on the left-hand side of a beam three divisions out.

a. Where could two squares be placed on the right to balance the beam?
b. Where could three squares be placed to balance it?
c. Where could one square be placed?

2. Three squares two divisions out can be balanced by two squares three divisions out. But that is not the only position where the two squares can be put on the right.
a. Suppose one of the squares on the right is put two divisions out, where must the other be put?
b. Suppose one of the squares on the right is put one division out, where must the other be put?

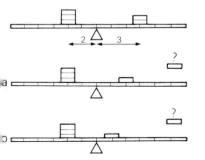

3. A pile of squares two divisions out on the left is balanced by four squares four divisions out on the right. How many squares are there in the pile on the left?

Experiment 9.2   Using a see-saw to measure your own mass

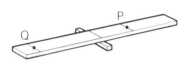

For this experiment you need a long, strong plank of wood. Balance it on a piece of wood. Mark a point P on one side of the plank which is 20 cm from the centre of the support. On the other side mark a point Q which is 80 cm from the support. Stand on the plank at P. Get someone to put 1 kg masses on the plank at the point Q until a balance is reached. Then you can calculate your mass.

If the plank is not long enough, or if you have not got enough 1 kg masses, you may have to choose different distances. We will leave that to you as an exercise.

The turning effect of a force

In experiment 9.2 the plank turns about the support in the middle: one side of the plank goes up and the other down. These experiments have shown that the tipping effect of a force depends on the size of the force and the distance of the force from the point about which the turning takes place. (This tipping or turning effect, the product of force × distance is sometimes called the *moment* of the force, but there is no need to memorise this name.)

More about forces

When two piles of squares balance each other as in the diagram, it is of course the gravitational forces acting on the masses which matter. The turning effect of the force P about the pivot is exactly equal to the turning effect of Q about the pivot.

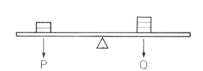

The forces *P* and *Q* are acting downwards, but the beam does not move. There must therefore be an upward force to balance them. Where is this force? It can only be at the

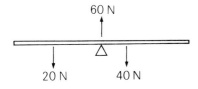

support and it will be equal to $(P+Q)$, but upwards, as in the example on the left.

The forces acting on a lever need not be gravitational forces. In the diagram below left, the force trying to turn the lever anti-clockwise is provided by a weak spring attached to the floor. In the diagram on the right the force is trying to turn the lever clockwise is provided by the load whilst the force trying to turn it anti-clockwise is provided by the weak spring pulling upwards.

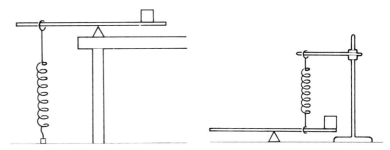

If a plank is balanced on a brick, there will be a downward force, the weight of the plank acting on it. As the plank does not move, it must be balanced by an upward force exerted by the brick. We can conveniently think of the weight of the plank acting downward through the balance point even though we know it is really distributed over the whole plank. If the plank is uniform (that is, it has the same thickness and width all the way along and the density is everywhere the same), then the balance point is in the middle. We call this balance point the *centre of gravity* of the plank.

Experiment 9.3   Finding the mass of a plank

Remember that we can think of the weight of the plank as acting at the centre of gravity and that this is at the centre of a uniform plank.

Move the plank so that it balances when a boy is standing on the end of it as shown in the drawing on the left. Measure the distance from the boy to the support. Measure the distance from the support to the centre of the plank. If you know the mass of the boy, you will be able to calculate the mass of the plank. (When you have done the calculation, try to check your result by some other means.)

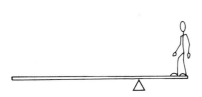

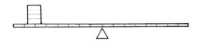

Questions for homework or class discussion

Four squares are placed six divisions out from the centre as shown on the left.

1. How many squares must be placed four divisions to the right to balance it? *6*

2. How many divisions to the right must three squares be put in order to balance it? *8*

3. How many squares must be placed two divisions to the right to balance it? *12*

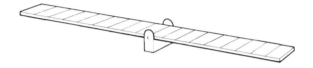

The see-saw shown above is balanced at its centre and equal divisions are marked on it outwards from the centre. Various masses are put on it. Say in each case whether they balance or not. If they do not balance, say which end will go down.

4. 10 kg four divisions out on the left; 20 kg eight divisions out on the right.

5. 10 kg four divisions out on the left; 20 kg two divisions out on the right. *balance*

6. 2 kg eight divisions out on the left; 4 kg four divisions out on the right. *bal*

7. 10 kg four divisions out on the left; 10 kg two divisions out and 20 kg one division out on the right. *bal*

8. 2 kg two divisions out and 4 kg five divisions out on the left; 8 kg two divisions out and 2 kg four divisions out on the right. *bal*

9. A plank of wood is balanced on a brick. A boy stands at a point 30 cm from the centre. Ten 1 kg masses are placed on the plank 1.5 m from the centre on the opposite side and the beam balances. What is the mass of the boy?

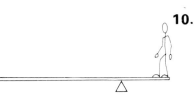

10. A heavy plank is 3 metres long. You can assume the centre of gravity is at the centre. If it balances on a support placed 0.5 m from the end when a man of 60 kg stands at the end, what is the mass of the plank?

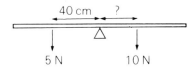

A light beam is supported at its centre and two forces of 5 newtons and 10 newtons act on it.

11. If the beam balances, at what distance from the centre must the 10 newton force act?

12. What is the upward force exerted by the support on the beam?

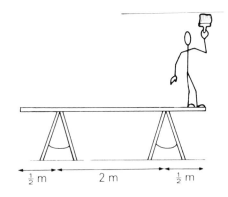

A painter stands on a plank of wood balanced on two trestles as shown. The painter has a mass of 80 kg. When he stands in the position shown, the plank starts to topple.

13. What must be the mass of the plank of wood?

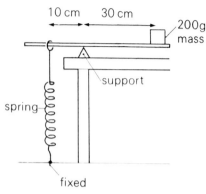

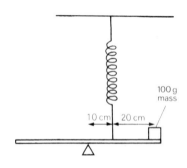

14. When the plank just starts to topple, the left-hand trestle exerts no force on the plank. What is the upward force exerted by the right-hand trestle on the plank? Assume that the gravitational force is 10 newtons on every kilogram.

A light beam is balanced as shown on the left by the force due to the spring acting on one side of the balance point and by a mass of 200 g on the other side. The gravitational force is 10 newtons on every kilogram.

15. What is the downward gravitational force acting on the 200 g mass?

16. What is the downward force that must be exerted by the spring?

17. What is the upward force exerted by the support?

The beam shown on the left is balanced by the upward force due to the spring and the downward gravitational force on the mass of 100 g in the positions shown.

18. If the gravitational force is 10 newtons on every kilogram, what is the downward force due to gravity on the 100 g mass?

19. What is the upward force exerted by the spring if the beam balances?

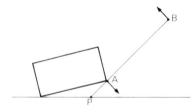

20. A crowbar is used to move a heavy crate as shown above. The crate exerts the force at A on the crowbar and this is just balanced by the force at B exerted by a man pushing the end of the crowbar. The distance AP is 10 cm and the distance PB is 200 cm. If the force at B is 40 newtons, what will be the force at A?

Revision test B1

1. The number of atoms in a gram of copper is
$$9\,500\,000\,000\,000\,000\,000\,000\,000.$$
Write this down using the power of ten notation. Give your answer in standard form.

2. How many grams are there in a kilogram?

3. Use your answers to questions 1 and 2 to write down the number of atoms in 1 kg of copper.

4. Express 178.248 4 correct to two decimal places. Then express it correct to one decimal place.

5. A man measures his height with a tape measure and he says it is 178.248 4 cm. What is unsatisfactory about this answer? What would be a reasonable value for him to have given for his height?

Revision test B2

1. The diagram shows a lever supported at its centre. Four metal squares of equal mass are placed at mark 4 on the left-hand side.
a. Can you balance the lever with eight more squares of the same mass, the original four squares remaining at mark 4 on the left? If your answer is yes, where would you put them? If your answer is no, give your reason.
b. With the four squares still on the left as illustrated, two squares are put on the right at mark 4. Where on the right would you put an additional pile of four squares to get a balance?

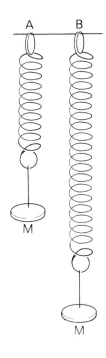

2. A mass M, hung on the end of a spiral spring A, causes that spring to stretch by 1.0 cm.

A new spring is made from similar wire and wound in exactly the same way as in spring A. Twice as much wire is used in spring B, so that B is twice as long as A when each is unstretched.

a. How much will B stretch if the mass M is hung on B?

b. If springs A and B are joined end to end to form one long spring, how much will the combined spring stretch when M is hung from the end? Explain how you get your answer.

3. Five boys are each given a ruler 1 metre long. They take it in turns to measure the length of their school drive from the gates to the front door. They return with the following results:

Smith	432 m 42 cm 1 mm
Jones	423
Roberts	43.0 km
Harrison	421 m
Brown	429.33 m

a. What is wrong with Jones's result?

b. In what way could you criticise the results of Smith and Brown?

c. Roberts made a mistake. Suggest what his mistake might have been.

d. What can you say about the length of their school drive?

4. Draw a labelled diagram of a microbalance.

a. How would you use it to compare the masses of two different postage stamps?

b. What does it mean when you make a balance *more sensitive*?

c. How can you make a microbalance more sensitive?

d. Would your microbalance still work if you took it to the Moon?

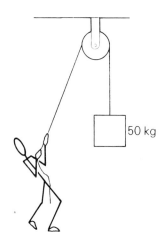

50 kg

5. A man pulls on a rope over a pulley as shown, and raises a mass of 50 kg.

a. If the gravitational force is 10 newtons per kilogram, what is the downward gravitational force on the 50 kg mass?

b. With what force must the man pull on the rope?

c. The pulley is fixed to the ceiling. When the man is pulling on the rope, what is the downward force on the ceiling?

Revision test B3

1. If a mass of 100 g is hung on the end of an 8 cm spring, the spring increases in length by 4 cm. When the mass is removed, the spring goes back to its original length.

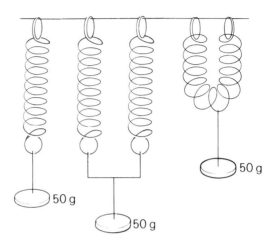

a. How much would you expect the length of the spring to increase, if a mass of 50 g were used?

b. A second similar spring is connected in parallel with the first spring (as in the middle drawing). How much would you expect each spring to increase now?

c. Instead of putting two springs in parallel, the original spring was bent double as shown in the diagram on the right. This is done carefully, so that the spring is disturbed as little as possible. The 50 g mass is hung from the middle. How far would you expect the 50 g mass to move downwards? Give the reason for your answer.

2. Draw a diagram to show how you could balance a 200 g lump of iron against a 300 g lump of lead, using a metre rule as a lever. Mark on your diagram possible values for the distances involved.

What must you do to make sure that the mass of the rule makes no difference?

3. You place a small block of expanded polystyrene measuring 3 cm × 4 cm × 5 cm on a lever-arm balance. The pointer measures 1 g as nearly as you can estimate. You are not fully convinced, since the balance tends to stick anyway, so you get hold of a big block of similar polystyrene measuring 30 cm × 40 cm × 10 cm. You put this on the balance, which reads 220 g. Use this to find a more accurate value for the mass of your original small polystyrene block.

4. Two lemon squash bottles have tight fitting corks. Both bottles appear to have nothing in them, but one is labelled 'air' and the other 'vacuum'. Their masses are measured on a balance and the bottle labelled 'vacuum' has a mass 2 grams greater than the other.
a. Is this a surprising result? Give a reason for your answer.
b. Describe how you would show there was a difference between the bottles, that one contained air and the other nothing.

5. Complete the following table.

Object	Volume/cm³	Mass/g	Density/g per cm³
P	100	?	2.5
Q	4	25	?
R	?	600	8
S	2 000	?	0.0012
T	?	8	0.25
U	3	24	?

Revision test B4

1. A pile of six pennies is put on a beam, 2 units from the balance point.
a. Can the balance be restored by putting a pile of three more pennies on the beam? yes
b. On which side would you put the three pennies? opposite
c. How far from the balance point would you put them?

2. A spring is fixed rigidly at its top end, while the lower end is loaded with various masses. The length of the spring is measured carefully with a ruler and the following table is obtained.

Mass/g	0	20	40	60	80	100	120
Length/cm	12	14	16		20	22	24

a. One reading was omitted by mistake. What would you expect to be the value which was left out in the table?
b. What would the length be if 85 g were the load?
c. Can you tell from the table what would be the length if 500 g were added? If the answer is yes, what would the length be? If the answer is no, explain why not.

3. The mass of an orange squash bottle is measured on a lever arm balance. It is then connected to a pump and the air is pumped out. The mass is again measured on the same balance and the reading is found to be the same as before.

Tom says this proves that air is weightless. Dick says it proves that air must have leaked back into the bottle. Harry says that it proves that the balance is not sensitive enough to show the difference. Which, if any, of these three statements do you think is true? Give your reasons.

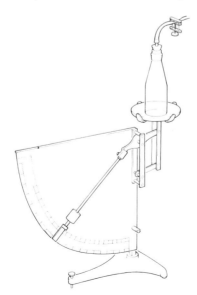

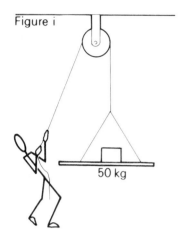

Figure i

50 kg

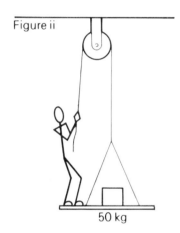

Figure ii

50 kg

4. Suppose you wish to measure the mass of a dead fly using a microbalance. Explain how you can make masses in order to calibrate the microbalance. How would you know what is their mass? Finally, explain how you would find the mass of the fly.

5. A platform has a load on it. The combined mass of the platform and load is 50 kg.

a. If the gravitational field has a strength of 10 N/kg, what is the downward force on the platform and load due to gravity?

b. What is the upward force on the platform due to the rope if the platform is at rest as in figure i?

c. What force is being exerted by the man on the rope in figure i?

d. The man has a mass of 80 kg. What is the downward force on him due to gravity?

e. The man stands on the platform as in figure ii and holds it in mid-air. What force must the man now exert on the rope to keep the platform still?

f. What will be the downward force exerted by the pulley on the ceiling in figure ii?

More estimates B5
Estimate the following.

1. The number of classroom lessons during a school year.

2. The number of peas in a pod.

3. The number of currants in a fruit cake.

4. The number of notes struck on a piano during the playing of a hymn.

5. The number of potatoes eaten by a boy in a year.

6. The number of blades of grass on a football field.

7. The number of chairs in your school.

8. The number of raindrops that fall into a jam jar during a heavy fall of rain.

9. The length of a bus in metres.

10. The number of stars you can see on a clear night.

Description of experiments B6

1. Describe what you would do to measure the distance between one groove and the next on a gramophone record.

2. How would you find the volume of a cork? What else do you need to know to calculate the density of cork? How do you work out the answer?

3. Suggest an experiment to find the density of an irregular piece of granite rock.

4. You have 50 small ball-bearing spheres. They are too small to measure on their own. Suggest how you would calculate their density.

5. How would you check the accuracy of a measuring cylinder which is marked in cubic centimetres if you have a rectangular transparent box and a ruler.

6. Someone breaks a newton spring balance in your laboratory. You agree to make a new one which can measure up to 10 newtons. Describe what you would do. You can assume you have a cardboard tube, a large variety of springs and various masses.

Chapter 11 **Pressure**

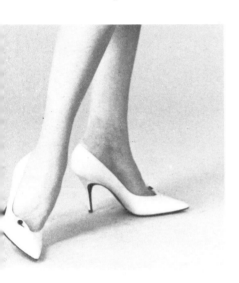

In earlier chapters, we have learnt about forces. We know it matters in what direction they are applied, and in the last chapter we learnt that it matters where they are applied. In this chapter we will learn that it sometimes makes a lot of difference over what area they are applied. An obvious example of this is the difference it makes when you sit on a drawing pin: if the flat side is uppermost you do not mind very much, but your feelings are very different if the point is uppermost.

The mass of the girl on the left might be 50 kg. She will therefore exert a force downwards on the floor of about 500 newtons. The mass of the elephant might be 1 000 kg. He will therefore exert a force downwards on the floor of about 10 000 newtons, which is 20 times greater than the force exerted by the girl. But when she walks on the wood floor with her stiletto heels she may do much more damage to the floor than the elephant.

When she walks, she may momentarily put all her weight on one of the heels and the area of the heel might be 1 cm². So a square centimetre of floor has to support her weight of 500 N. When the elephant walks, he may momentarily put all his weight on two feet, but their combined area might be about 500 cm². He is therefore exerting a force of 10 000 newtons on 500 cm², or 20 newtons on each square centimetre. This is much less than the 500 newtons on one square centimetre of floor exerted by the girl.

Experiment 11.1 The effect of area

Take some modelling clay. It needs to be reasonably soft for this experiment, so work it in your hands beforehand and get it pliable. Put a layer of it, at least 2–3 cm thick in the pan of the scales. (It would be better to use scales which have a scale measured in newtons as they will then tell you the

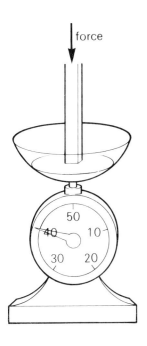

force

force applied, but it may be necessary to use ordinary scales used for measuring mass.)

For the actual experiment, you will need a piece of wood 1 cm × 1 cm in cross-section, another 2 cm × 2 cm and preferably some other sizes as well. Hold the rod over the scales and push it down on to the modelling clay until the reading on the scales is, say, 40 newtons (or 4 kg if they are scales for measuring mass). Hold the rod at that reading for a short while.

Repeat the experiment, using a rod with a different area of cross-section.

Push with the rod on a fresh part of the modelling clay until the reading on the scales is the same as before. Repeat with pieces of wood with other areas of cross-section.

Examine the clay to see how deep a dent was made by each of the rods.

The force was exactly the same each time, but the deepest dent was made when the area was smallest. When the force was spread over a large area, the dent was small.

Pressure

To explain these differences, we use the word *pressure*, defining it as the force divided by the area.

In the above experiment, the force in each case was 40 newtons. The first rod had an area of 1 cm², so the pressure was $\frac{40}{1}$ or 40 N/cm². The second rod had an area of 2 cm × 2 cm, or 4 cm². The pressure was $\frac{40}{4}$ or 10 N/cm². This pressure was less than before and the dent was therefore less.

$$\text{PRESSURE} = \frac{\text{FORCE}}{\text{AREA}}$$

To make dents of the same depth with both rods we should have needed to apply a force of 160 N to the 2 cm × 2 cm rod.

Questions for class discussion

1. What was the pressure exerted by the girl in high heel shoes at the beginning of this chapter? What was the pressure exerted by the elephant? Which would dent a wooden floor the most?

2. Make an estimate of your mass. What is the downward gravitational force on you in newtons? Estimate the area of the flat top of a drawing pin. What is the pressure when you sit on the flat side of the drawing pin? Estimate the area of the pointed tip of a drawing pin. What is the pressure if you sit on that, assuming that all your weight is supported by the point?

3. If a boy falls through some ice on a pond, why is it better for a man to crawl across the ice when he goes to the rescue? The best thing for the man to do is to put a plank of wood, or a ladder on the ice and to crawl along that. Why is that?

4. Skis are very awkward things to carry around. Why is it necessary to have them so large?

5. Why do grocers use a fine wire for cutting cheese?

6. You can carry a suitcase comfortably by the handle. But if you have no handle and you tie the case with string, why is it very painful carrying the case with your hand through the string?

Pressure of liquids

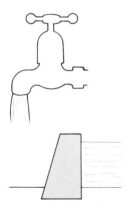

So far, we have considered the pressure due to solids alone, but liquids also exert a pressure. You can feel it for yourself by turning on a tap and putting your thumb over the outlet. You might try this with a tap at the top of a house and then with a tap at ground level or in a basement. Do you notice any difference? (You may not notice any difference if the taps provide drinking water from the mains. It should be very noticeable with the hot taps.)

Water pressure is obviously very important when building a dam or the wall of a reservoir. The water will exert a considerable pressure on the wall. You will notice in the drawing on the left that the wall is thicker at the bottom than it is at the top. Why do you think this is?

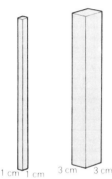

Questions for class discussion

Imagine a column of water with height 30 cm and cross-section 2 cm × 2 cm.

1. What is the volume of the water?

2. The density of water is 1 g/cm³. What is the mass of the water?

3. The gravitational field has strength 10 N/kg. What is the downward force due to gravity on this mass of water?

4. What is the area of the base on which this force acts?

5. What is the pressure on the base, that is, the force on 1 cm²?

6. In just the same way, work out what the pressures would have been if the base had been 1 cm × 1 cm or 3 cm × 3 cm.

7. What do you notice about the results?

8. What would happen to the pressure if the height of the column were 60 cm and not 30 cm?

9. What would happen to the pressure if the density were 2 g/cm³ and not 1 g/cm³?

These examples have shown that the pressure does not depend on the area of the base: the pressure was the same in each of the cases in Questions 5 and 6. It also showed that the pressure depends on the height: if you double the height, you double the pressure; if you treble the height, you treble the pressure and so on. Similarly it depends on the density: if you double the density, you double the pressure.

So far, we have considered the pressure on the base. But does a liquid exert a pressure in any other direction?

Experiment 11.2   Investigating the direction and strength of water pressure

Take a tin can and make holes in it with a round nail. It helps when making these to put a block of wood inside as an anvil. Put the holes at different places round the can but at the same level. Take care to make the holes the same size. Fill the can with water and watch how the water spouts out. If you wish, you can put the can under a tap so that water flows in at the same rate as it flows out.

Secondly, take another can and make equal holes, one near the bottom, one near the top and one in the middle. Watch how the water flows out of these holes.

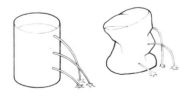

Finally, take a third can and batter it into an irregular shape. Make holes in three or four different places and watch what happens this time.

This experiment should confirm that the pressure is certainly greater at the bottom than it is near the top. You will also notice that the water comes out at right angles to the surface. Of course it does not continue in that direction because gravity acts on the jet and it falls towards the earth, but it always starts at right angles to the surface.

Optional experiment for the bath

Fill a balloon or polythene bag with water. Make some holes in it with a pin (one near the top, one in the middle, one near the bottom) and watch what happens. The jets should all start at right angles to the surface: the jet from the one near the top will start upwards.

Experiment 11.3   Water finds its own level

It is often said that water always finds its own level'. What does this mean? And what is the cause of it?

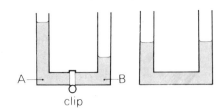
A——B
clip

Two glass tubes should be held vertically and joined at the bottom by rubber or polythene tubing. Put a clip in the middle so that water cannot flow from one side to the other. Put some coloured water in the left-hand tube so that it is nearly full (a few drops of ink will colour the water) and put less coloured water in the right-hand tube.

We know that the pressure depends on the height of the water. It is therefore greater at A than it is at B. But A and B are at the same level. In consequence, when the clip is opened, water will flow from A to B. It will continue to flow until the pressures are the same, in other words until the water levels are the same.

Suppose the two tubes were unequal in size as shown below. What would happen in these cases when the clip is opened?

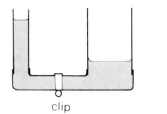
clip

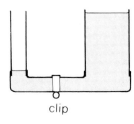
clip

From the above we see it is not the mass of water on the two sides that must be the same. It is merely the pressure that is the same and that means that the heights must be the same. The shape of the tubes does not make any difference as is shown in the tea-pot below, or the odd shaped glass container.

 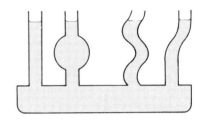

Homework assignments

1. Read the topic book *Pressures*, pages 16–18, and prepare a short talk on non-return valves. You should mention the valves of a bicycle tyre and of a motor car tyre.

2. Read *Pressures*, page 22, and prepare a talk on the public water supply.

3. Find out what you can about the hydraulic brake system in a car and write a short essay on it. (See *Pressures*, pages 27–29.)

Pressure of gases

Gases also exert a pressure. If you put your finger over the outlet of a bicycle pump and push the handle in, you will feel the force produced by pressure. If you blow up a balloon with a small hole in it, the air will come out of the hole whichever direction the hole is pointing. What about area? The next experiment will tell us something about that.

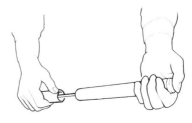

Experiment 11.4 Feeling forces with two syringes of different sizes connected together

Hold the small syringe while pushing on the larger one so that air goes from one syringe to the other. Feel the forces on the two syringes. Then push on the small syringe while holding the large syringe. Feel the forces again. The force is greater on the plunger with the larger area than it is on the plunger with the smaller area.

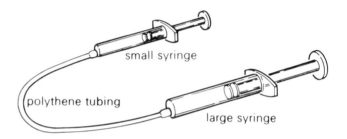

small syringe

polythene tubing

large syringe

You can then repeat the experiment with the syringes and connecting tube filled with water. Feel the forces as the water is pushed from one syringe to the other. The pressure of the water is the same everywhere inside, but the forces are different because of the different areas.

Experiment 11.5   The pressure box

The apparatus shown below has a plastic bag inside it. There are two moveable platforms resting on the top, one is four times the area of the other. Close the outlet tube and blow air into the inlet tube until the bag is full of air.

Place a 1 kg mass on each platform and increase the air pressure inside by blowing into the box. Which of the platforms rises first? Why does this happen?

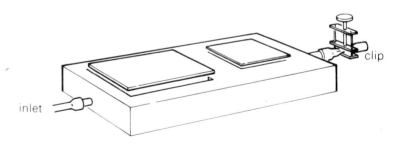

clip

inlet

Repeat the experiment, keeping one 1 kg mass on the small platform but with two, three and four 1 kg masses on the large platform. Which rises first on each occasion? Do these results agree with your knowledge of pressure?

Atmospheric pressure

Earlier in the course, you measured the density of air. 1 cm³ of air has a mass of 0.001 2 g which is not very great, but our atmosphere contains a great deal of air extending upwards for many kilometres. We are in fact at the bottom of a great sea of air and this exerts a large pressure on us, about 10 newtons per square centimetre. We are so used to this pressure that we do not ordinarily notice it, but it is there none the less. The following demonstration experiments should convince you of its existence.

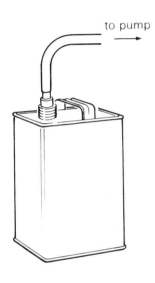

to pump

Experiment 11.6   Collapsing can
A can, with a well fitting bung is connected by pressure tubing to a vacuum pump. At first the pressure of the air inside the can is the same as the atmospheric pressure pressing on the outside of the can. The pump is then used to draw the air out of the can and the can collapses because of the atmospheric pressure on the outside.

Experiment 11.7   Demonstration of atmospheric pressure
Rubber sheet is tied over the open end of a bell-jar (so called because it is shaped like a bell). The other end of the jar has a well fitting bung and is connected by tubing to a vacuum pump. At first the atmospheric pressure on one side of the sheet is balanced by the pressure due to the air inside the bell-jar. The pump then takes away the air inside and the effect of the atmospheric pressure is seen.

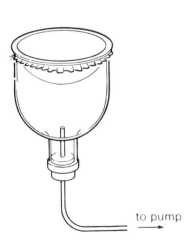

to pump

Experiment 11.8   Balloon in a bell jar
Put a partially filled balloon inside a bell jar which is placed with its open side on a thick glass plate. Some vacuum grease is probably necessary to seal it and so make it airtight. Connect the top of the jar to a vacuum pump. At first the atmospheric pressure on the outside of the balloon

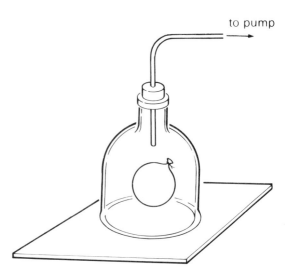

balances the air pressure inside, but when the vacuum pump removes the air and there is no longer the same pressure on the outside, the balloon expands.

Experiment 11.9   Another collapsing can

For this experiment you need a can with a good bung or air-tight cap. Put a little water in the bottom of the can and boil it vigorously with the top open. The water vapour will drive much of the air out of the can. Then close the top. Let the can cool. The water vapour inside will condense back to liquid. The pressure inside will be much less than the atmospheric pressure outside and the can will collapse. (If you try this experiment at home, it is wise to use a rectangular can. Some cylindrical cans will stand up to big pressure differences and will not collapse.)

Class discussion about further experiments

1. A tumbler is filled with water and a piece of paper is put on the top. It is important that there are no air bubbles in the water. With your hand holding the paper on the tumbler, turn it over and remove your hand. Why does the water stay in the glass?

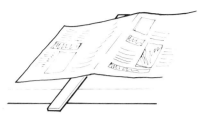

2. A sheet of newspaper is placed over a thin piece of wood on a table. If a small steady force is applied to the end of the wood, what happens? If you give a sudden sharp blow to the wood, what happens? Explain the difference.

3. A drinking straw is filled with liquid. Place a finger over both ends and hold the straw vertical. When you take your finger off the bottom, the liquid stays in the tube. Why is this?

4. If you make a single hole in a lemonade or beer can and turn it upside down, the liquid does not flow out. Why are two holes necessary?

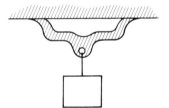

5. A sucker pressed on to a flat surface can support quite a heavy mass. Why is this?

6. If a piece of cotton wool or cloth is wedged in the bottom of a tumbler of water and the tumbler is pushed down below the surface of the water in a deep bucket, the wool does not get wet. Why is this?

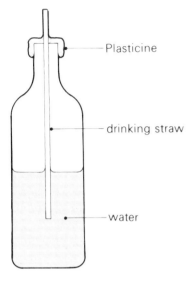

Plasticine

drinking straw

water

7. Take a milk bottle and fill it with water. Put a drinking straw in it. Fix Plasticine round the drinking straw at the top of the bottle and make sure you have an air-tight joint. Can you suck the water out of the bottle? Explain what happens.

8. Using the same bottle, drinking straw and Plasticine, see how many air bubbles you can blow through the drinking straw. When you have got as much air as possible into the bottle, take your mouth away from the drinking straw and see what happens. (It is wise to do this near a sink!) Explain what occurs.

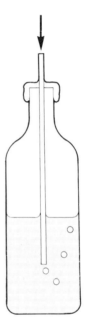

Manometers

A useful device for measuring pressure differences is a manometer. This is a U-shaped tube of glass or clear plastic, partially filled with liquid.

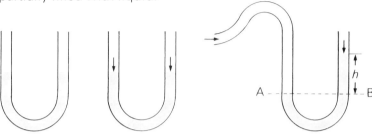

If the tube is open at both ends, as shown above, the level of the liquid will be the same on both sides. The pressure of the atmosphere is of course pressing down on the liquid surfaces, but as it is the same on both sides the levels are the same. If you raise the pressure on one side by blowing into a tube attached to that side, as shown in the third diagram, the levels will be different. When the liquid is at rest, the pressure at A will be the same as the pressure at B. The pressure at B will be equal to the pressure of the atmosphere plus the pressure due to the column of liquid, h. The excess of pressure on the left is therefore measured by the height, h.

A series of experiments with manometers will help you to get familiar with them. It is suggested that the following should be set up as a circus of experiments. You can try each in turn, spending about ten minutes with each.

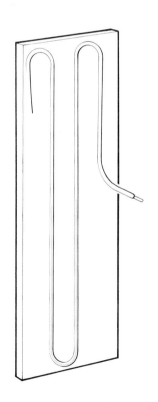

Experiment 11.10 Measuring your lung pressure

For this experiment you need a manometer almost $2\frac{1}{2}$ metres high, half filled with coloured water. Blow hard into one side of the manometer and when you have got the water level on the other side as high as possible, get your partner to note the water levels. Measure the distance between the two levels. What is your lung pressure in centimetres of water?

Then try sucking, in other words see how much you can lower the pressure. Can you suck as big a difference as you can blow?

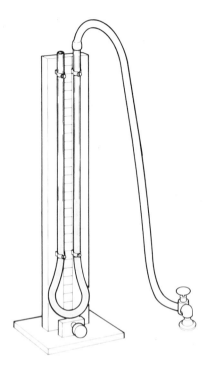

Experiment 11.11 Measuring the pressure of the gas supply

Connect the manometer to the gas supply. Turn on the gas supply and leave it on. Measure the distance between the water levels. What is the gas pressure measured in centimetres of water?

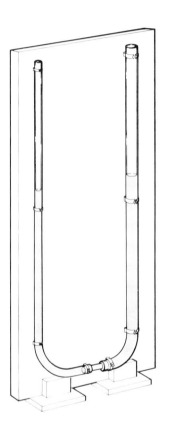

Experiment 11.12 Using a manometer with unequal tubes

In this experiment use a manometer in which one of the tubes is much wider than the other. Join the narrower tube to the gas supply and measure the difference in levels. Then join instead the wider tube to the supply and measure the difference in levels. What do you find? Is this what you would expect?

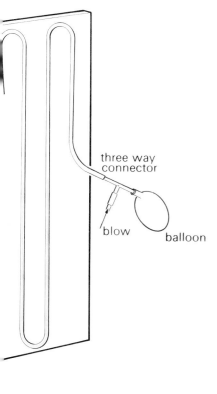

three way
connector

blow   balloon

Experiment 11.13 Comparing a water manometer with a mercury manometer

A water manometer and a mercury manometer are connected through a 3-way connector to a bicycle pump with a non-return valve. Use the bicycle pump to increase the pressure inside the tube. How do the levels compare in the two manometers? The density of mercury is almost fourteen times the density of water. Does this explain the difference?

Experiment 11.14 Measuring the pressure necessary to blow up a balloon

A balloon is tied securely to a 3-way connector, which is connected to a large manometer as shown on the left. Blow up the balloon through a tube connected to the 3-way connector. Get your partner to watch how the pressure changes. Does it stay the same as the balloon is blown up?

The Bourdon gauge

Another device for measuring pressure is the Bourdon gauge. You are doubtless familiar with the toy, shown on the left, used at parties. The harder you blow (in other words, the greater pressure you exert) the more it uncurls. The Bourdon gauge works on the same principle, the greater the pressure applied the more the curved tube inside straightens out and this moves a pointer across a scale.

Bourdon gauges can be calibrated to measure the pressure directly in N/cm^2.

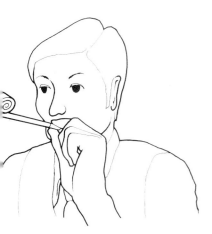

Experiment 11.15   Using a Bourdon gauge to measure pressure

Connect a tube to the Bourdon gauge and blow into it to measure your lung pressure. Also try sucking to see how low a pressure you can reach.

Experiment 11.16   Using a Bourdon gauge to measure the pressure of a balloon as it is blown up

This is a repeat of experiment 11.14, but this time using a Bourdon gauge instead of a manometer.

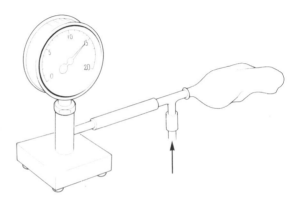

Experiment 11.17   Bourdon gauge to measure low pressures

Connect a Bourdon gauge direct to a vacuum pump. Switch on the pump and see how low is the pressure you can get. Great care must be taken in this experiment when letting air back in again: if it rushes in too fast, it can damage the gauge.

Measuring atmospheric pressure

Atmospheric pressure, usually about $10 \, \text{N/cm}^2$, varies from day to day and from place to place. It plays a very important part in predicting the weather and for that reason it is necessary to be able to measure it.

Experiment 11.18   Height of a mercury column

A glass tube, 1 metre long, is held vertically with its lower end below the surface of some mercury in a trough. (The trough should be in a tray in case any of the mercury gets spilt.) The top end of the tube is connected to a vacuum pump. As the pump might be seriously damaged if mercury got into it, it is necessary to include a round-bottomed flask as shown in the drawing as a trap for any mercury.

The mercury in the tube will start at the same level as the mercury in the trough as the pressure both outside and inside the tube is atmospheric.

If the pump is switched on for a short while, the air pressure in the tube will be lowered and some mercury will rise up the tube.

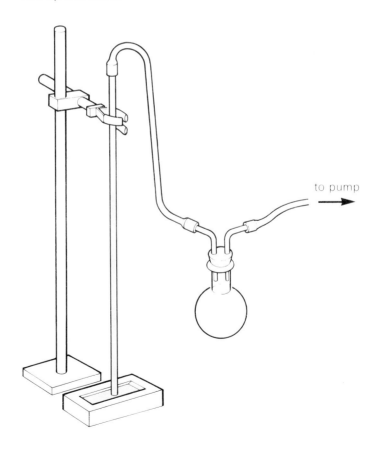

to pump

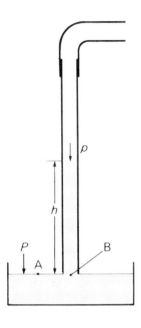

The pressure at A is the atmospheric pressure, P, due to atmosphere pushing down on the mercury in the trough. If the mercury is not moving, the pressure at B must be the same as the pressure at A. The pressure at B, however, will be equal to the pressure, p, due to the air left at the top of the tube plus the pressure, h, due to the column of mercury.

If the pump is switched on again, more air will be taken away, the pressure, p, will be less and the mercury in the tube will rise further.

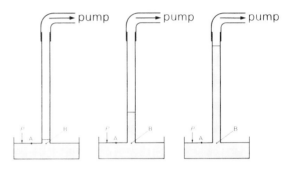

Will it be possible to draw the mercury right up to the top of the tube? That might be serious as we have said it is possible to damage a pump if mercury gets into it. Let us see what happens when you pump continuously.

You will find that the mercury rises to a height of almost 76 cm and then will rise no more.

The pressure at A is the atmospheric pressure and this is the same as the pressure at B. Assuming the pump is a good one, the pressure at the top of the tube is now almost zero. (It would be zero if there were no air at the top of the tube and there was a perfect vacuum there.) So the pressure at B is merely the pressure due to a column of mercury 76 cm high.

This is the reason why you will sometimes find pressures measured in centimetres of mercury. If you read that the pressure is 76 cm of mercury, it means that the pressure is the same as that exerted by a column of mercury 76 cm high.

Barometers

It is possible to measure the atmospheric pressure more simply by setting up a barometer, as in the following experiment.

Demonstration experiment 11.19   Setting up a simple barometer tube

The barometer tube is usually made of thick glass for strength, and it is sealed at one end. First, it must be completely filled with mercury: this is done over a tray in case of spillage. A finger is then placed over the end of the full tube and the lower end is held below the surface in a trough of mercury. (You must wash your hands afterwards unless you wore rubber gloves. Mercury is a dangerous substance.) Can any air get into the barometer tube when the finger is taken away? Clearly there is no way in which it can.

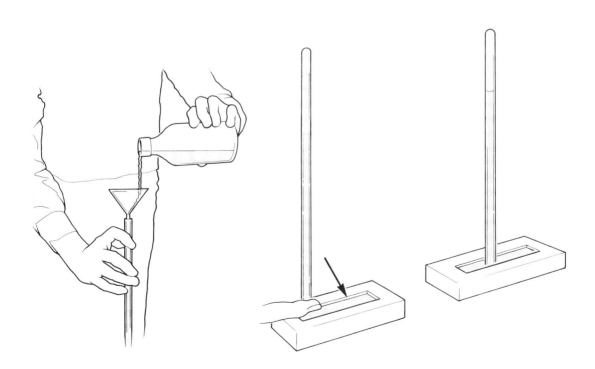

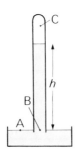

When the finger is removed, the mercury level in the tube will fall until the height, h, is such that the pressure at B, due to the column of mercury, is the same as the atmospheric pressure at A.

As no air got into the tube, what will there be in the top of the tube at C?

You can now measure the atmospheric pressure in centimetres of mercury by measuring the height, h.

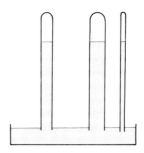

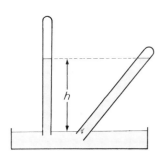

Would it make any difference to the height, h, if a barometer tube with a larger diameter were used? To find the answer to this, set up two barometer tubes of different diameter side by side. What you already know about the pressure due to a column of liquid should tell you the answer.

Finally, what happens to the height, h, if the barometer tube is inclined at an angle? Again find the answer by trying it, although once again you should be able to predict the answer.

A water barometer

The height of a mercury barometer is about 76 cm. Mercury is 13.6 times as dense as water. What would be the height of a water barometer? In other words, what would be the height of a column of water which would exert the same pressure as the atmosphere?

As an optional project, you might like to set up a water barometer in your school. You will need tubing which is at least 11 metres long, though it would be perfectly all right to join a number of lengths together provided the joints are air-tight. You might like to use glass tubing for the top section so that you can see the level. You will need to fill the tube with water.

A useful tip might be to fill the tubing with water while it is on the floor. When it is full put a bung into the top of the tube and leave the lower end in a deep bucket of water: a water butt might be wiser in case anything goes wrong. And of course, you will need somewhere high to set it up, perhaps outside a building, or the well of a staircase. Such a barometer was set up by Otto von Guericke outside his home in Magdeburg in 1657, and the photograph shows one set up in a school.

Water barometer at St. Hugh's School, Woodhall Spa

Questions for homework or class discussion

1. A boy has a mass of 50 kg. The gravitational field exerts a force of 10 newtons on each kilogram. If the total area of his shoes in contact with the ground is 100 cm², what is the pressure exerted on the ground?

2. A car has four tyres at an excess pressure of 15 N/cm². The mass of the car is 1 000 kg. What is the weight of the car in newtons? What area of each tyre is in direct contact with the ground?

3. A car has a mass of 1 200 kg. It is supported by four tyres which have a pressure of 20 N/cm². After several weeks use, the pressure falls to 16 N/cm². How is it possible for this lower pressure to support the car?

4. Is there a maximum length for a drinking straw used for drinking milk out of a bottle?

5. A gas tap is connected to a manometer containing mercury. When the tap is turned on, the pressure difference is 1 cm of mercury.

a. What would be the difference in levels if a water manometer were used? The density of mercury is 13.6 g/cm³.

b. If the difference in levels is 3.4 cm when a manometer with another liquid in it is used, what would be the density of this liquid?

6. This question is about the way in which a syphon works. Initially the tap T is closed.

a. What will be the pressure at A?

b. B is a short distance above A. Will the pressure at B be less than, equal to or greater than the pressure at A?

c. What will be the pressure at C?

d. D is a large distance above C. Will the pressure at D be less than, equal to or greater than the pressure at C?

e. Which will be greater: the pressure at B or the pressure at D?

f. Will water flow from B to D, or from D to B, when the tap T is opened?

g. At what stage will the water cease to flow from the upper bowl to the lower one?

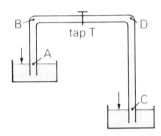

7. To fill a syringe, the plunger is pushed in and the end of the syringe is put in the liquid. As the plunger is pulled out, liquid enters the syringe. Explain why this happens.

8. The inside of aeroplanes are usually 'pressurised'. What does this mean? Why is it necessary to do this pressurising?

9. If the atmospheric pressure is 10 N/cm², what will be the total force due to the atmosphere on a flat roof which measures 12 m × 9 m?

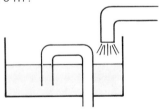

10. The above is a drawing of an automatic flushing cistern. Water flows into the tank from the tap which is left running permanently. When will it start to flush? When it has started, at what stage will it stop flushing?

11. On a certain day the pressure due to the atmosphere is equivalent to the pressure exerted by a column of mercury 75 cm high.

  a. The density of mercury is 13.6 times the density of water. What would be the height of a column of water which exerted the same pressure as the atmosphere?

  b. The density of water is about 1 000 times greater than the density of air. If we assume that the density of air is always the same, what would be the height of a column of air which would give the same pressure as the atmosphere?

  c. This gives a possible figure for the height of the atmosphere. In fact, the height of the atmosphere is greater than this. Has anything gone wrong in the calculation? Explain why the actual height is greater.

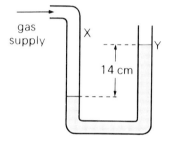

12. When a water manometer is connected to a domestic gas supply the difference in levels is 14 cm.

What would be the difference in the levels of water in X and Y if:

  a. tube Y were twice as wide as tube X?

  b. tube Y were half as wide as tube X?

  c. mercury were used instead of water? (Suppose mercury is 14 times as dense as water.)

    Could an astronaut on the Moon use this method to measure the pressure of his oxygen supply? Give reasons for your answer.

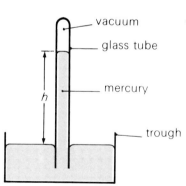

13. The diagram shows a barometer.

  a. What, roughly, is the distance marked h?

  b. What does this tell us about the atmosphere?

  c. Give a reason for using mercury rather than water.

  d. If we pushed the tube 2 cm down into the trough, what would happen to the distance h?

  e. What would happen if there were a very small hole in the glass at the top of the tube?

14. What does a barometer measure?

How, if at all, would the reading of a mercury barometer be altered if:

a. the tube were to have double the diameter?

b. the barometer were taken to the top of a mountain? Give a reason.

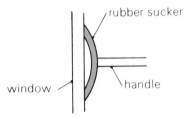

15. A rubber sucker is moistened and pressed against a window thereby pushing out most of the air inside the sucker.

a. Explain why it is difficult to remove the sucker from the window.

b. Why was it preferable to moisten the rubber sucker?

16. This is a diagram of a simple lift pump for raising water out of a well. A non-return valve is a regulator which allows fluids to flow through it one way but not the other. A and B are non-return valves.

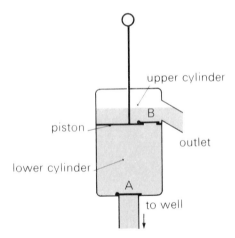

a. Describe the action of the valves and water when the piston is pushed down from its present position.

b. Describe what happens when the piston is raised again.

c. What makes the water come up the tube out of the well?

d. Why can it not take water out of a well more than 10 m below the pump?

Homework assignments

1. Read about the aneroid barometer and the barograph (*Pressures*, pages 35–36), and either write an essay on them or prepare a short talk.

2. Read about syphons in *Pressures*, pages 24–26, and find out about the water flushing system in your lavatory. Give a short talk on how it works and include an explanation of the use of a ball-cock.

3. Write an essay on pumps, including in it details of how a lift pump and a force pump work.

4. In 1657 Otto von Guericke conducted a famous experiment in the city of Magdeburg in which he tried to pull apart two hemispheres using two teams, each of eight horses. Find out what you can about this experiment and write an account of it.

5. A knowledge of atmospheric pressure is very important in predicting weather. Find out what you can about weather maps and prepare a short talk on them. It will make the talk more interesting if you collect some samples of weather maps from newspapers.

Chapter 12 Particle Model of Matter

In Chapter 2 we looked at crystals, we noticed the regularities of shape and we saw what happened when crystals grew. We found we could begin to understand this if we imagined that matter was made up of particles.

We found support for the model when we saw crystals of calcite being cleaved. If the regular shape of crystals meant that they were made up of layers of particles, we might expect them to cleave along certain planes and that is just what happened.

Models play a very important part in science. You cannot ever *prove* that a model is *correct*. All you can do is to collect more and more evidence which makes it likely. You go on believing it is a good model until you find evidence that contradicts it. Then you have either to abandon the model or to modify it to fit the new evidence. But models do not have to be correct in all respects to be useful, provided their limitations are remembered. We found a model of polystyrene spheres useful, but it certainly does not mean that all matter is made of little polystyrene spheres!

Solids, liquids and gases

In this chapter we will extend our model further, and in particular we will see how it can be applied to liquids and gases. To do this we start by thinking about what happens to substances when they are heated.

Experiment 12.1 Turning solids to liquids and liquids into gases

a. Put some ice in a beaker. Stand the beaker on a tripod over a bunsen flame and watch what happens to the ice.

b. Put a little water in a tin lid, heat it until the water boils. Then go on heating. What happens to the water?

c. Hold a bunsen burner at an angle of 45° so that the flame is directly over a mat of asbestos. Hold the end of a short piece of solder in the flame. The flame should be held at an angle so that the molten solder falls on the asbestos and not into the burner. What happens to the solder?

d. Hold a short length of lead strip in the flame, with the bunsen burner still at an angle, and watch it melt. What happens if a piece of iron wire and a piece of copper wire are held in the flame? (It is probably best to hold the pieces of wire in a pair of pliers. Why do you think this is a good idea?)

e. Put some sulphur in a deep crucible and heat it in a flame. Take care to see the sulphur does not catch fire (that is why it is wise to use a deep crucible). See what happens to the molten sulphur when it cools. Do you see crystals forming?

f. Put some naphthalene in a test tube to a depth of about 2 cm. Then hold the test tube in a beaker of boiling water. What happens to the naphthalene? It is better to melt naphthalene in hot water rather than hold it over a bunsen flame since it gives an unpleasant smoke if it catches fire. Watch what happens to the molten naphthalene when it cools. Do you see crystals forming?

You know that ice melts when it is warmed and that if you boil water it turns to steam. You will have seen other examples of solids melting and turning to liquids in the experiments above, although you were not able to melt iron wire in the bunsen flame. Iron does not melt until it reaches a much higher temperature. The melting points of various substances are given in the list below.

copper	1 083 °C	lead	327 °C	silver	961 °C
ice	0 °C	mercury	− 39 °C	sulphur	113 °C
iron	1 535 °C	naphthalene	77 °C		

Heating liquids causes them to turn to gas, as happens when water turns to steam. The boiling points of some substances are given below.

copper	2 595 °C	mercury	357 °C	silver	2 212 °C
iron	3 027 °C	methylated spirit	79 °C	sulphur	445 °C
lead	1 744 °C			water	100 °C

If substances which are gases at room temperature are cooled, it will be found that they turn to liquid and further cooling will turn them into solids. For example, oxygen at normal pressure liquefies at -183 °C, nitrogen at -196 °C, hydrogen at -253 °C. They turn from liquid to solid at still lower temperatures: oxygen -219 °C, nitrogen -210 °C, hydrogen -259 °C.

Molecules and atoms

So far we have referred to a solid as being made of particles. We ought now to consider what these particles are.

A lot of important evidence about the particles of which matter is made comes from chemistry. It tells us that matter is generally made up of *molecules*. A crowd is composed of people, a library is composed of books, a forest is composed of trees. Matter is composed of molecules. A molecule is the smallest part of the substance which is still that substance. It is possible to cut up trees or tear up books, but to do so changes the tree or the book. It is the same with molecules. Molecules can be broken into *atoms*; for example, a molecule of water can be broken up into two hydrogen atoms and one oxygen atom, but if so it becomes hydrogen and oxygen and it is no longer water.

Chemistry tells us that there are many different kinds of atom which occur in nature. These are the *elements* of which hydrogen and oxygen are two. Other elements of which you will have heard are nitrogen, carbon, lead, copper, zinc, iron, sulphur, and so on.

Quite often atoms do not like going around alone and they prefer to be in pairs. For example, hydrogen atoms usually pair off together and so do oxygen atoms. A molecule of hydrogen therefore consists of two hydrogen atoms, and a molecule of oxygen of two oxygen atoms.

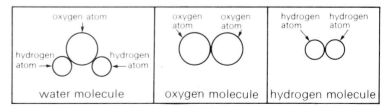

| water molecule | oxygen molecule | hydrogen molecule |

The drawings represent the atoms of oxygen and hydrogen as spheres. This is merely another example of a model which we find very convenient. It does not mean that they are necessarily spheres. We have to go a lot further in our study before we can say more precisely what an atom is like, and even then we will merely be replacing one model by another.

Questions for class discussion on the particle model of a solid

1. The study of crystals led us to a model of a solid made of particles. If you take a piece of metal and pull it, it is very difficult to stretch it. What does this tell you about the forces between the particles?

2. It is very difficult to squash a piece of metal. What does this tell you about the forces between the particles?

3. (A harder question.) If you take two pieces of metal and hold them close to each other, they are not attracted to each other. Furthermore, if you touch the two pieces together, they do not stick to each other. What does this tell you about the forces between the particles?

4. If you pull a piece of steel wire so that it stretches a little and then let go, it returns to the original length. What does this tell you about the forces between the particles?

5. (Much harder.) If you pull on a piece of copper wire so that it stretches, it does not always go back to the original length. How can you explain this?

The particle model

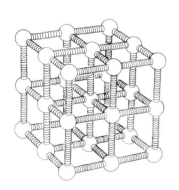

We can now form a model of a solid made of particles with strong forces between them. When we try to pull the particles apart, there are forces of attraction pulling the particles together. When we try to squash the model, there are repulsive forces pushing them apart. Normally the forces of attraction and repulsion balance each other. A convenient model for our solid is shown on the left, in which polystyrene spheres are joined by springs.

The particles are all fixed in position, but they can vibrate. When heat is added, they vibrate more and more but the general arrangement remains the same. If more heat were added, the particles might vibrate so much that some of the links between them would break, and so the solid would turn into a liquid. In a liquid there are still forces between the particles, but they are not as strong as before and the liquid is therefore able to flow.

In the previous chapter, we learnt about the pressure exerted by a gas. Perhaps now we can explain this in terms of our model. If heat is given to a liquid, the molecules may be able to break away from each other entirely and, if sufficient heat is given to the liquid, all the molecules will be freed and will move around as a gas. If the gas is in a container, the moving molecules will bounce against the wall and will exert a force on it (in the same way that a ball bounced against a wall exerts a force on the wall). It is this force which is the cause of the pressure we have already discussed.

The particle model of gases and the use of the model to describe their properties is usually referred to as the *kinetic theory* of gases: kinetic comes from a Greek word meaning motion. Some further models may help to illustrate it.

Experiment 12.2 Two dimensional kinetic model

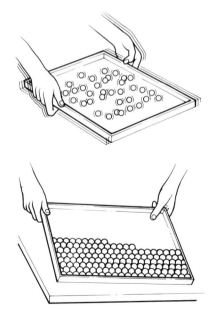

Put about twenty coloured marbles in the tray provided, and place the tray flat on the table. Agitate the tray, keeping it on the table, with an irregular shaking movement. Watch the marbles moving about.

Try to use marbles which are of similar colour and have one which is distinctive and different, say red. Watch the motion of the red marble as the tray is agitated.

Listen to the collisions that occur. You should be able to hear two kinds: collisions when a marble hits a wall and collisions between the marbles themselves. The collisions with a wall cause a force to be exerted on it. Perhaps the pressure exerted by a gas is due to the collisions with the walls.

Tilt the tray and agitate it in the tilted position with most of the marbles at one side. Why do you think this gives a possible model for a liquid?

Questions for class discussion

1. What path does a single marble take when the tray is being agitated?

2. Is the length of the path taken by one marble between collisions with other marbles always the same, or does it vary?

3. If, in the above experiment, you agitate the tray rather more violently (in other words, 'heat' the 'molecules' of the gas so that they move faster), what do you notice about the collisions with the walls? (From this model, we might expect the pressure exerted by a gas would be greater if the molecules were travelling faster because there would be more impacts per second and each impact would have more effect. And this is exactly what we do find.)

4. If some more marbles are put in the tray, the model now represents a gas with a greater density. How does this affect the collisions with the wall if the tray is agitated so that the marbles have the same speed as before? Does this agree with the fact that the pressure of a gas does increase with density if the temperature is kept constant?

Experiment 12.3 Partitioned two-dimensional model

Put marbles into a tray with wooden partitions fixed in it as shown. The three different sections represent possible models for solid, liquid and gas. Agitate the tray and notice the difference in the behaviour of the marbles in each section. Do you think this is a good model?

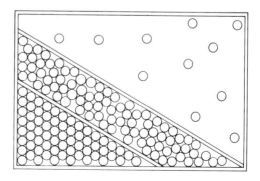

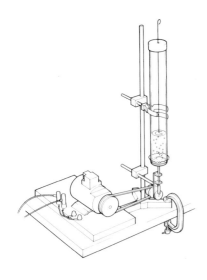

Experiment 12.4   Three-dimensional model

In a gas, the molecules can move in all directions so perhaps a better model is provided by the apparatus on the left in which the small ball-bearings in the plastic tube represent the molecules of a gas which move in three dimensions. An electric motor drives the vibrating rod under the rubber base and this sets the ball-bearings in motion. A cap on the top of the plastic tube prevents any of the balls escaping.

If the speed of the motor is gradually increased, the ball-bearings will fly around faster and faster, colliding both with each other and with the walls of the tube.

If a loose-fitting cardboard disc is lowered inside the tube, it will be bombarded by the ball-bearings. It will settle in a position when the downward force on the disc due to gravity (its weight) is balanced by the average upward force due to the collision of the balls on the disc.

What happens if an extra cardboard disc is added to the top of the other disc?

Support for the particle model: change in volume

If the kinetic theory of matter is a good one, we ought to find some more evidence to support it. In our model of a solid the atoms or molecules are very close together, but in a gas they are a long way apart. We should therefore expect that the density of solids and liquids would, on the whole, be much greater than the densities of gases. That is exactly what we found earlier: the density of aluminium was 2.7 g/cm³, copper 8.9 g/cm³, water 1.0 g/cm³, methylated spirit 0.79 g/cm³, but the density of air was 0.001 2 g/cm³.

This merely compares the densities of different substances. What would you expect to happen to the volume of a substance when it changes from liquid to gas? The following experiment shows what happens.

Experiment 12.5   Change of volume when water changes to steam

This experiment requires the use of a large syringe and a small hypodermic syringe. Push in the piston of the large syringe so that the volume reads zero and then fit a rubber cap over the end. Put the syringe in a deep beaker of brine and bring it nearly to boiling. (Brine is merely water with common salt dissolved in it, but it has the property that it boils at a temperature a little above 100 °C.) Then partially fill the hypodermic syringe with water. Take the large syringe out of the water and inject into the rubber cap exactly 0.1 cm³ of water. When the needle is withdrawn from the cap it seals up so that neither water nor steam can escape through the hole. Now put the large syringe into boiling brine.

The water in the syringe will turn to steam and the steam will push out the piston showing that there is a very large increase in volume. You will probably find that the 0.1 cm³ of water produces over 100 cm³ of steam giving a change in volume of 1 to 1 000. The accepted value for the change in volume is about 1 to 1 600.

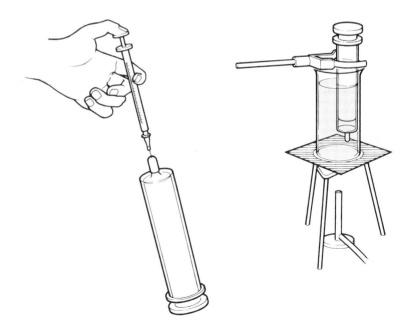

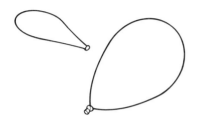

Experiment 12.6 Change of volume when 'dry ice' turns to gas (demonstration)

This experiment is possible only if your teacher can get some solid carbon dioxide, 'dry ice'.

Two people should hold open the neck of a rubber balloon. Put into the balloon a spoonful or two of solid carbon dioxide. Flatten the balloon quickly and tie a firm knot in the neck. What happens to the solid carbon dioxide as it warms up?

The change in volume is about 1 to 600. This experiment is also interesting since it shows a solid turning straight into gas without going through a liquid stage.

Support for the particle model: Brownian motion

There is a very important experiment which supports the idea that a gas consists of molecules moving randomly in all directions.

The trouble is, that molecules are much too small for us to see, even with a very powerful microscope, and in any case we think they would all be moving very fast, but in this experiment we can see a direct result of these fast moving molecules.

To understand the experiment, imagine a lot of people all sitting round a football which is suspended by a string from the ceiling. Suppose they start throwing marbles at the ball. When it is hit by a marble it would move, though not very much because it is so much more massive than the marble. Sometimes it would be hit on one side, then on another side and so on. The football would move first in one direction, then in another. This random irregular motion of the football would be a direct result of the collisions.

In the experiment, you will look at smoke particles in a little cell containing air. You cannot see the molecules of the air, but you can see the light scattered by the smoke particles as bright points of light. The movements of these show the irregular motion of the smoke particles as each one jiggles around when the air molecules bump into it. But before you look at the actual experiment, two other experiments will help you to understand it.

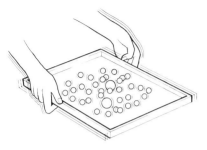

Experiment 12.7 Two-dimensional model for Brownian motion

For this experiment, use the same apparatus as in Experiment 12.2. This time, put a much larger marble in the centre of the tray. Once again agitate the tray when it is flat on a table and watch the heavier marble. You will notice it has a much slower irregular motion.

Then put in the tray a piece of expanded polystyrene. It can be any shape as long as its base is flat. Watch the polystyrene as the marbles knock it around. It moves irregularly.

Experiment 12.8 Three-dimensional model for Brownian motion

This time, use again the same apparatus as in Experiment 12.4. Put an expanded polystyrene sphere (about 1 cm in diameter) among the small ball-bearings. Set the vibrator in motion and watch the polystyrene sphere being knocked around in an irregular motion by the ball-bearings as they collide with it.

We should, however, be careful. These experiments do not tell us anything about gases since they are only models. The ball-bearings or the marbles are *not* molecules of gas; they merely represent the molecules in our model. To get real evidence about the molecules, we must do the actual Brownian motion experiment.

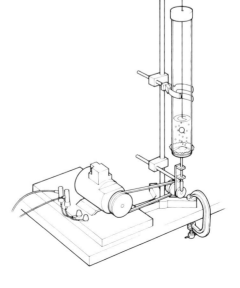

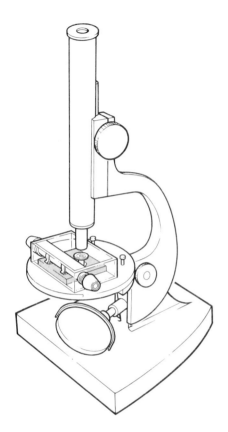

Experiment 12.9   Brownian motion in a smoke cell

The apparatus consists of a small glass cell which you are going to look into under a low-powered microscope. The smoke particles must be illuminated. So the apparatus includes a small lamp and a glass rod which acts as a lens to concentrate the light on the middle of the cell.

First, remove the cover from the cell and fill the cell with smoke. The simplest method is to hold a drinking straw almost vertically over the cell. Then light the *top* end of the straw so that the smoke pours down the inside of the straw and into the cell. When it has plenty of smoke in it, put back the cover to prevent the smoke escaping and place the apparatus on the platform of the low-powered microscope.

Connect a 12-volt supply to the lamp so that the light shines on the cell. Focus the microscope on the cell until you can see the light scattered by the smoke particles as bright points of light in it. Watch for a while and you will see the irregular motion. This irregular motion is due to the air molecules hitting the smoke particles even though we cannot see the molecules. (The smoke particles are *not* smoke molecules, they are tiny specks of soot.) This is exciting, for it is the first time you have seen direct evidence for the existence of molecules.

Questions for class discussion

1. When you look at the smoke particles, some of the specks will suddenly become rather larger patches of light. What is the reason for this?

2. Sometimes one of the specks of light will disappear altogether. What has happened to the smoke particle?

3. What would be the difference if rather larger and more massive smoke particles were used?

4. If the temperature were raised, we think the air molecules might move faster. How would this affect the motion of the smoke particles? What might happen if the air were cooled?

5. How do you know that the motion of the smoke particles is not due to the light shining on them?

Brownian motion is so called because it was first observed by a Scottish botanist called Robert Brown. He observed it when watching pollen grains in water. The motion was due to the pollen grains being knocked around in an irregular way by the water molecules. It is possible to see the Brownian motion of carbon particles in water if a very small speck of graphite (Aquadag or Indian ink) is added to a few cubic centimetres of distilled water. Light should be shone through the water, and the particles will have to be viewed with a higher powered microscope than was necessary in Experiment 12.9. You might like to try this if you have time.

Support for the particle model: diffusion

Support for the particle model also comes from diffusion. You will have noticed that if your mother cooks fish and chips in the kitchen and leaves the door open, it is not long before the smell has reached all over the house. Does this support the idea that gases consist of particles in constant and rapid motion? If your teacher puts a small dish of ether in one corner of your laboratory, how long is it before you can smell it everywhere else in the room? Further examples of diffusion in gases and liquids are given in the following experiments.

Experiment 12.10　Diffusion of nitrogen dioxide into air

You can prepare nitrogen dioxide gas by putting a mixture of equal volumes of concentrated nitric acid and water on some copper turnings at the bottom of a gas jar. The gas produced has a distinctive brown colour. When the action has stopped and the gas has cooled to room temperature, invert another gas jar containing air over the top of it. Watch what happens. It will help if you hold a sheet of white paper behind the jars.

Does this show that molecules of nitrogen dioxide have diffused into air? Is there any evidence that air has also diffused into the nitrogen dioxide?

This experiment was done with the brown nitrogen dioxide underneath. (Nitrogen dioxide is denser than air and that seemed the sensible way in which to do the experiment.) What might have happened if the gas jar of air were underneath and the gas jar of nitrogen dioxide on top? Try it and see. This will show that diffusion occurs as before.

Experiment 12.11　Speed of molecules diffusing in air

When molecules of hydrochloric acid meet with molecules of ammonium hydroxide, they react and form a white powder or smoke. In this experiment a few drops of hydrochloric acid are put at one end of a glass tube while a few drops of ammonium hydroxide are put at the other end. A rubber stopper is inserted at each end of the tube.

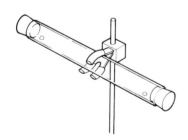

Notice how long it takes before any smoke appears. Notice also where the smoke first appears. What does this tell you about the speeds at which hydrochloric acid molecules and ammonium hydroxide molecules diffuse through air?

There may appear at first sight to be some contradiction here. The Brownian motion experiment suggested that the air molecules were moving very fast and in the above experiment it took a few minutes for the hydrochloric acid molecules to diffuse down the tube. But there is a simple explanation for this. When the hydrochloric acid molecule diffuses through the air, it does not travel in a direct straight line. Its motion is an irregular one in which it is repeatedly

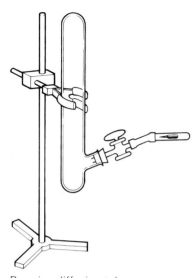

Bromine diffusion tube

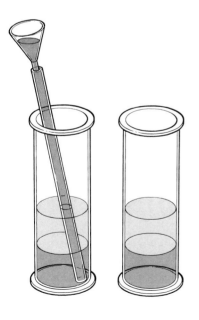

colliding with air molecules. Its path will be much like the path of the single marble you watched in the two-dimensional kinetic model (see page 128).

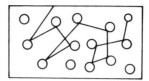 

A man can run at speed from one end of a street to another if the street is empty. But if the street is full of people, all moving in different directions, and if he runs at the same speed he will be buffeted all over the place in an irregular way. It will take him much longer to travel the same distance. Molecules are like that. They travel at the same speed in both cases, but get buffeted about by air molecules and take longer to move down the tube when air molecules are also present.

At a later stage of your course you will probably see an experiment in which some bromine gas is released at the bottom of a glass tube containing air. You will notice that it takes time for the brown bromine gas to diffuse through the air. You will then see another experiment in which the glass tube is connected to a vacuum pump and the air is removed from it. Bromine gas is then released into a vacuum and the result is dramatic. The bromine molecules can now rush straight through the tube without hitting air molecules on their way. One can well believe in molecular speeds of hundreds of metres per second!

Experiment 12.12 Diffusion of copper sulphate solution into water

Put water at the bottom of a gas jar. Add a concentrated solution of copper sulphate very slowly down a funnel and tube. The copper sulphate is more dense than water and will go to the bottom. Great care should be taken so that the liquids do not mix: there ought to be a distinct line of separation between the liquids. Diffusion in liquids will be seen after a few hours.

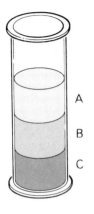

A

B

C

Another version of this experiment uses three different liquids. Put a strong sugar solution in the bottom of the glass jar. Put the concentrated copper sulphate solution on top of that. Finally, add water. After an hour or two, you will notice that copper sulphate molecules will have diffused both up into the water and down into the sugar solution.

Experiment 12.13   Diffusion of potassium chromate into gelatine

Mix some gelatine and hot water (about 50 g of gelatine in 400 cm³ of water). Put the mixture into four boiling tubes: fill one of them, and half-fill the other three. Leave the gelatine to set in the tubes.

Fill up the last three tubes with potassium chromate solution and close the tubes with rubber bungs to exclude any air bubbles. Stand the tubes so that one is upright, one upside down, one on its side and the one full of jelly put anyhow. Leave them for a day or two and then look to see if diffusion has occurred into the jelly. Does it make any difference which way up the tube is? The fourth tube which was completely filled with gelatine is often referred to as 'a control'. What do you think is the reason for that? Does the gelatine in it get any of the potassium chromate colour when it is left?

After doing the experiment, open the tubes and replace the potassium chromate solution by pure water. Leave the tubes again and in due course you will see diffusion of the coloured potassium chromate out of the gelatine into the water. Diffusion appears to take place in all directions.

A summary

All the evidence in this chapter has supported a model of matter made up of very small particles or molecules.

In a solid the molecules are held in a regular array. They can vibrate to and fro, but do not move around. There are strong short-range forces which make it hard to pull the molecules apart, and there are also strong short-range repulsions which make it difficult to compress a solid.

In a liquid, the molecules are still fairly close together (there is not much change in volume when a solid melts), but their motion is not restricted to vibration and they can move around freely. A liquid does not have a definite shape, but the forces keep the molecules sufficiently together to ensure that the liquid has a definite volume.

In a gas, the molecules are much further apart. The forces of attraction between them are now very small, perhaps zero, and so they move around freely filling the space available. They have an irregular motion and travel with high speeds. They exert a pressure on the walls of the container when they collide with them.

The size of molecules and atoms

How small are atoms or molecules? So far we have no idea except that they must be much smaller than the smoke particles we used in the Brownian motion experiment. You will shortly do an experiment in which you will estimate the size of a molecule, but first we must look at the skin effect of liquids (usually referred to as Surface Tension).

Experiment 12.14　Experiments on surface tension

a. Look at the shape of drops of water forming on the end of a slowly dripping tap. Try to watch the shape of the drops as they fall.

b. Look at the shape of small drops of mercury on a glass surface. Compare them with the shape of drops of water on the glass surface. Put a little wax from a lighted candle on to the glass and then put a drop of water on the wax. What is the shape of the drop this time?

c. Make a frame of wire about 5 cm in diameter. Dip the frame into a dish containing a soap solution (50 % water, 50 % liquid detergent can be used) so that a soap film is formed on the frame.

Tie a short length of cotton to form a continuous loop. Put it in the soap film as shown. Then puncture the film inside the loop using a piece of chalk. What is the shape of the loop now?

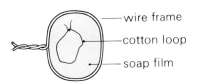
wire frame
cotton loop
soap film

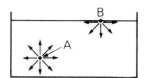

These experiments all suggest that the surface of a liquid behaves as though it had an elastic skin trying to keep the area as small as possible. There is not really a skin, but the liquids behave as if there were.

This *surface tension* effect can be explained as follows. The molecules of the liquid attract each other (unless they are very close indeed when they repel). For a molecule A inside the liquid, the attractive forces pull in all directions. But a molecule B at the surface will have no liquid molecules above it to attract it. It will therefore tend to be pulled into the liquid and this means that the surface tends to shrink whenever possible. It therefore behaves as though it had a thin elastic skin.

Questions for class discussion

1.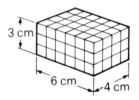

Blocks of wood, all 1 cm cubes, are piled up to form a stack which is 6 cm long, 4 cm wide and 3 cm high. How many blocks are there in the pile? What is the volume of the pile in cubic centimetres?

2.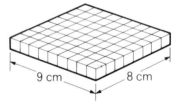

The pile of cubes in Question 1 is spread out to form an array 9 cm long by 8 cm wide, but only 1 cm high. Calculate the new volume. Is this the result you would expect? How does the volume compare with the answer in Question 1?

3. A pile of cubes (each with side 1 cm) measures 10 cm long, 8 cm wide and 6 cm high. How many cubes are there? What is the total volume?

   The cubes are spread out so that they form a layer 1 cm thick. What is the area covered by the cubes?

4. A piece of butter measures 2 cm × 3 cm × 4 cm. What is the volume of the butter in cubic centimetres?

The butter is spread out into a slab which has everywhere the same thickness. The slab measures 4 cm by 6 cm. What is the volume of the butter? What is the area of the slab? What is its thickness?

5. Suppose the slab of butter in Question 4 is rolled out further until it covers an area of 240 cm². What is its thickness now, assuming it is everywhere the same?

6. You pour lead shot from a small beaker on to a tray so that it lies in a pile. You then spread it out into a flat layer. What is the thinnest layer you can get?

7. Suppose the volume of the lead shot in Question 6 is 100 cm³. If the area covered by the lead shot is 100 cm², what is the thickness of the layer of lead shot? Suppose the lead is spread out into a layer one lead shot thick, and the area covered is now 1 000 cm². What is the size of the lead shot?

8. If you did the above experiment spreading out lead shot until it was one layer thick, the result might be rather like the photograph below. There are no forces of attraction between the lead shot and there would probably be empty patches.

Lead shot in tray

But if instead of lead shot you spread some oil on top of a water surface, the oil would float on the water and spread out. The surface tension forces would however keep the oil film together. What would be the smallest possible thickness for the oil film?

141

How would the volume of the oil film compare with the original volume of the oil? This should give you a clue how we are going to measure the length of an oil molecule.

Measuring an oil molecule

How small are atoms? If we lined up atoms side by side, how many would there be in 1 cm? We have already seen that scientists like to make estimates, but in this case it is a bit difficult to guess. Might it be 1 000? Or a million? Or a million million? We really have no idea, so it would be good if we could make some measurement which gives us any idea of the size. We shall choose a molecule of olive oil, which is in the shape of a long chain of atoms, about twelve atoms long. Unfortunately it is much too small for us to measure directly, but we will try to measure it by a roundabout method and it will be a great achievement if we can get some answer.

Experiment 12.15   Preliminary experiments
In all these experiments it is extremely important that everything is very clean. Thoroughly wash the crystallizing-dish with detergent and then rinse it several times in clean water to remove all the detergent.

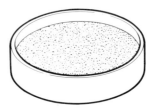

Put water in the dish and lightly sprinkle the water surface with lycopodium powder. Put a drop of alcohol on the powdered surface and watch what happens. (If very little happens, or nothing at all, it means that the dish was not properly cleaned.)

A clear patch appears in the powder as the alcohol spreads over the surface. After a while, the powder partially returns as the alcohol either dissolves in the water or evaporates.

Dip the end of a matchstick in some olive oil and wipe it

clean. Then put the end of the matchstick in the water surface. Again watch the clear patch appear as the oil spreads over the surface. This time the oil film remains and the powder does not return as it did when the alcohol was used.

Finally, clean the dish again with detergent and water. Dust the surface again with lycopodium powder. Then dip a clean finger in the surface. Perhaps you will be surprised at the amount of oil on an apparently clean finger: that is why it was necessary to get everything so clean.

Now we are ready for the actual experiment. The previous experiment will have shown you how oil spreads on a surface. There have been examples of oil spreading on the sea when a leak has occurred in an oil tanker and the oil patch has spread over a very large area.

In this experiment we will take a very small drop of oil, measure its volume and then let it spread on a water surface. We will measure the area of the oil film and hence we will calculate the thickness of the oil film.

The chemists tell us that the olive oil molecule is such that one end is strongly attracted to water and the other is not. The oil molecules are therefore upright on the water, much like the bristles in a brush. Our experiment will therefore give us the length of the oil molecule assuming the oil film is only one molecule thick.

Experiment 12.16   Oil film experiment

For this experiment, a tray should be filled to the brim with clean water and dusted lightly with lycopodium powder.

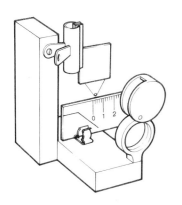

It is necessary to prepare a drop of oil which is $\frac{1}{2}$ mm across. Put a transparent scale marked in $\frac{1}{2}$ mm (see left) in the holder provided and in front of it fix a hand lens. Dip the loop of steel wire in olive oil and take up a drop. Fix the card in the holder so that the drop and $\frac{1}{2}$ mm scale are clearly seen through the lens. Use a second loop dipped in the oil to 'tickle' the first drop until it appears to be $\frac{1}{2}$ mm across. It is often helpful to use the second loop to run small drops together until you have one of the right size.

Carefully carry the $\frac{1}{2}$ mm drop to the tray and lower it into the centre of the water surface. It will immediately spread in a circle over the surface. Measure the maximum diameter of the film using a metre rule.

Calculating the size of the oil molecule

a. The drop was $\frac{1}{2}$ mm across. What is this measured in centimetres?

b. It will make the calculation easier if we assume the drop is a cube. We know the volume of a cube is the length multiplied by the breadth multiplied by the height. What is the volume of the oil drop? Give your answer in cubic centimetres.

c. When the drop is put on the water surface it spreads out to form a film one molecule thick. What is the volume of the film in cubic centimetres?

d. But the volume of the film is also equal to the length of the film × the breadth of the film × the thickness. Again to keep the arithmetic simple, let us suppose the oil film was a square patch. You have measured how many centimetres it was across, so that you know the length and the breadth. You can calculate the area and then you should be able to work out the thickness of the film. When you have done this, you will have made your first measurement on the atomic scale.

The chemists tell us that the olive oil molecule consists of twelve carbon atoms along its length, so that if you divide your answer by twelve, you should get the size of a carbon atom. Of course, there were approximations in your calculation and it was not easy to be very accurate in your measurements, but it is a fine achievement to have got an answer of the right sort of magnitude.

The generally accepted value for this is a little more than

$$\frac{1}{100\,000\,000}\text{cm} \quad \text{or} \quad \frac{1}{10^8}\text{cm}.$$

This means that in a centimetre, there might be approximately 100 million atoms lying side by side.

Chapter 13

Energy

What is energy?

You have often heard phrases like 'feeling energetic' or 'being full of energy', and there are many advertisements for food which 'give you energy'. What is meant by this word *energy*?

If you say that you have no energy, you mean that you do not feel like doing anything. If you have a lot of energy, then you can do a lot of jobs. You need energy to lift a pile of books, you need energy to run round a field or to ride a bicycle, you need energy to saw wood or to hammer nails. We all get the necessary energy for this from the food we eat: the advertisements are certainly correct in that breakfast cereals do give us energy. Food provides us with energy, which is stored up in chemicals in our bodies and this energy enables us to do jobs: we call this *chemical energy*.

If you lift a lot of bricks, run up and down stairs a lot of times or play a hard game of football, you are tired at the end of it. You feel hungry and you must have some more food (*chemical energy*) before you can do much more.

Jobs of work can also be done by engines of various kinds. Lorries can carry loads up a hill, cranes can lift masses, electric motors can drive saws. Do these engines need energy? Of course they do. A car will not go if it does not have petrol. A diesel engine requires oil. An electric iron and a washing machine need a supply of electricity to do the jobs for which they were made. A gas fire or a gas cooker needs a supply of gas. In the same way that human beings need to be fed, engines need to be fed with fuel. Energy is stored not only in food, but also in petrol, diesel oil, wood, coal, coke and gas. We say that all of these have a store of *chemical energy* in them.

Questions for class discussion

1. Think about each of the following jobs. Which of them require fuel and which need none?

a. Lifting a pile of books on to a shelf.

b. Watching a pile of books on a shelf.

c. Kicking a football.

d. Hitting a post into the ground.

e. A post holding up a clothes line.

f. The sea keeping a boat afloat.

g. A sailing boat moving across the sea.

h. An ocean liner crossing the Atlantic.

i. Sleeping in bed.

j. Reading a book.

k. Climbing a mountain.

l. Winding a clock.

m. A train travelling from station to station.

n. Holding up a pile of books.

2. Make a list of five things different from those above which require fuel and five things which do not.

Uphill energy

Look at a brick lying on the floor. Has it got any energy? Can it do a job of work? It does not seem very likely.

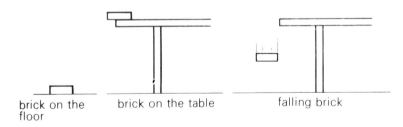

brick on the floor   brick on the table   falling brick

But if you pick it up and put it on the edge of a table, has it got any energy? The brick looks much the same, but if it falls from the table to the ground, it is capable of doing a job of work. It could certainly knock in a nail placed on the floor or it could make a dent in your foot.

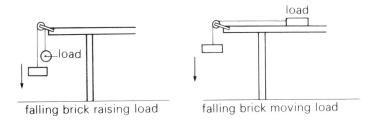

falling brick raising load   falling brick moving load

Suppose the raised brick is attached to a mass over a pulley as shown above. The falling brick can lift the mass off the floor if the mass of the brick is great enough. In the second diagram, the falling brick can pull the load along the table top. In both these ways the raised brick is capable of doing a job of work; in other words it has energy. But when the brick has reached the floor, it seems to have lost that energy.

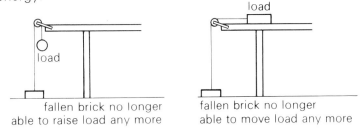

fallen brick no longer able to raise load any more   fallen brick no longer able to move load any more

Since this energy depends on the height of the brick, we will call it *uphill energy*. The higher the brick the more uphill energy it has got. Of course to raise it, we have to use some of the chemical energy stored in our muscles. What we have done is to transfer that chemical energy to the uphill energy stored in the brick.

Later in the course (page 237) we will call this energy *gravitational potential energy* or more briefly *potential energy*. But at this stage it is probably clearer if we call it uphill energy.

Motion energy

Put a thin plank of wood on the floor at the foot of a table, as shown. (You may prefer to use balsa wood or hardboard.)

Stumps knocked over by a cricket ball

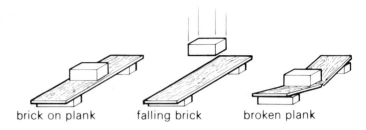

brick on plank   falling brick   broken plank

Put the brick on the wood. It is not able to break it.

Now raise up the brick to the top of the table so that it has got uphill energy. Then gently push the brick off the edge of the table. It falls on to the wood and this time it does break it.

The brick had uphill energy when level with the table, but this was almost gone when it reached the wood. None the less it was able to do the job of breaking the wood. It must therefore have some kind of energy if it can do a job of work. Since it is energy due to the fact that it is moving, we will call this energy *motion energy*.

In a bowling alley, it is *motion energy* that causes the pins to be knocked over; a moving cricket ball can knock over the stumps; a moving car can do a lot of damage if it hits something. Obviously the faster the body is moving, the more motion energy it has got. (Later we will call this *kinetic energy*, but at this stage it will be easier just to call it *motion energy*.)

Pins being knocked over in a bowling alley
Smashed car after a collision

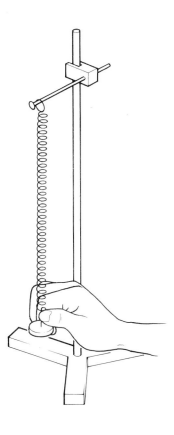

Spring energy (strain energy)

Fix the top end of a spring and pull down on the other end so that the spring is stretched. If you attach a mass to the bottom of the spring and let go, the spring will do a job of work and will raise the mass. In other words the stretched spring must have energy stored in it. We will call this *spring energy* or *strain energy*.

Further examples of strain energy can be seen in the drawings below. The stretched catapult, the arched bow, the bent branch of the tree all have *strain energy* stored in them; they are all capable of doing a job of work.

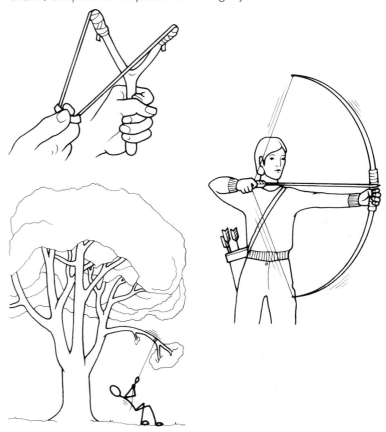

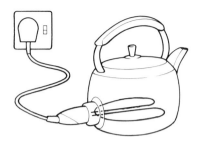

Electrical energy

Electric trains can do a useful job moving people and goods from one place to another. Electric motors can be used to lift masses or make all sorts of devices work. Electric fires will warm a room, electric kettles will boil water. In each case energy is needed to do the jobs and this is obtained from the electric supply. We will call this *electrical energy*.

Electric train

Energy transfer

A brick on the ground has no energy. If you lift it up and put it on the edge of a table, it has some uphill energy. In order to lift it, you had to use some of the chemical energy stored in your body: if you spend a lot of time lifting bricks you certainly feel hungry at the end. Energy has been transferred from one form to another. Chemical energy in the food was transferred to chemical energy in your body, which was transferred to the uphill energy in the brick. If the brick is then allowed to fall, the uphill energy will transfer to motion energy just before the brick hits the ground. This is shown in the following diagram.

food	CHEMICAL ENERGY	human body	UPHILL ENERGY	brick	MOTION ENERGY

Diagrams like these are very helpful for showing energy transfers and you will see many more examples later in this book.

Heat energy

If you take a block of wood with a flat surface and place it on a plank of wood, there is no rise in temperature of the wood however long you leave it. It makes no difference if you increase the forces between the block and the plank by loading the block.

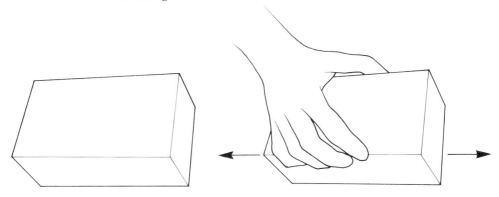

Instead of this, take the block in your hand and rub it a short distance backwards and forwards on the plank. If you do this vigorously for a minute or two, pressing down hard as you rub, you will find that both the block and the plank become warm. (A good way to feel this is to hold the block against your cheek before you start rubbing and then to touch your cheek again after rubbing.) Some heat has come from somewhere. Do you perhaps feel a bit tired after doing the rubbing? Have you used up some chemical energy?

Rub your hands together vigorously. You are using some of the chemical energy in your body to give your hands some motion energy. What do you notice after doing this? Perhaps that you are a bit tired and certainly your hands will feel hotter. Some of the chemical energy has been transferred to heat.

We know that burning coal, wood or paraffin oil produces

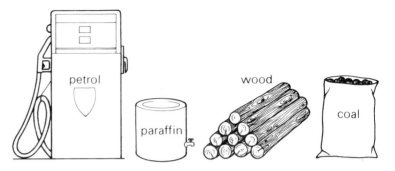

petrol

paraffin

wood

coal

heat. The chemical energy in those fuels has been turned directly into heat. In fact we believe that heat is another form of energy.

When the brick falls off the table, we have already seen that the uphill energy turns to motion energy as it falls. Just before it hits the ground, all the uphill energy has become motion energy. When the brick hits the ground, what happens to this motion energy? We believe that it turns to heat energy: both the brick and the ground will be a little hotter after the collision. The following flow diagram shows how the energy is being transferred.

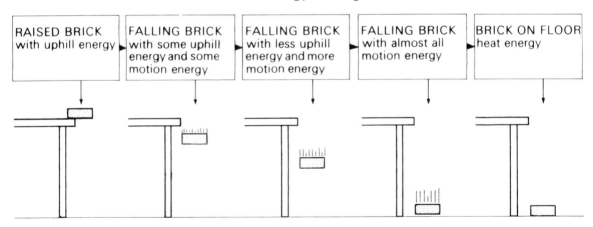

| RAISED BRICK with uphill energy | FALLING BRICK with some uphill energy and some motion energy | FALLING BRICK with less uphill energy and more motion energy | FALLING BRICK with almost all motion energy | BRICK ON FLOOR heat energy |

Heat caused by friction

It is important to remember that friction is only a force, and forces on their own do not cause a transfer of energy. To show this, put a block of wood on a rough plank. If you tilt

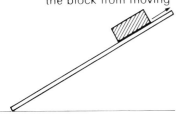

the plank, the block does not slide down (provided the slope is not too great). It is stopped from sliding down by friction, the frictional force between the block and the plank. But however long you leave the block in this position, it does not get hot. The frictional force does not cause any heating.

Earlier you rubbed a block of wood over a plank and the block became hotter. You were moving the block against the frictional force. There has to be movement against a frictional force for energy to be transferred to heat. If your bicycle is not moving and you apply the brakes, nothing gets hot. But when the brakes are applied to a moving wheel, the brake blocks can get very hot: the motion energy in the bicycle is transferred to heat energy.

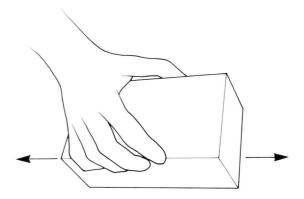

When a ball flies through the air, there is a small frictional force on it due to the air. As the ball is moving against this force, the air and the ball will be warmed a little. Some of the motion energy of the ball is transferred to heating the air and the ball, and so the ball is slowed down slightly.

In most transformations some energy gets transferred to heating the surroundings. For this reason, the flow diagram for a batsman hitting a ball would be more correct if we drew it like this.

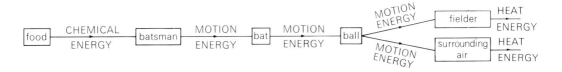

Lathe

Questions for homework or class discussion

1. A ball is at rest at the top of a hill. It starts to roll down it, getting faster the further it rolls. At the bottom of the hill, it hits a wall and comes to rest. Draw an energy flow diagram for this.

2. When an electric drill is used to drill a hole in wood or metal, the end of the drill becomes very hot. Where has the heat come from? List three other examples of energy being transformed into heat.

3. When a lathe is used in a metal workshop, there is usually a flow of liquid on to the cutting tool. What is the reason for this?

4.

In the drawings above, an archer takes a bow and bends it with an arrow in place. The arrow flies through the air and finally enters the target. Draw a flow diagram showing the energy transfer starting from the chemical energy in the archer's body.

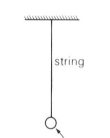

string

pendulum bob

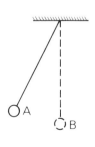

5. The drawing on the left shows a simple pendulum hanging at rest. In that position it does not have any energy. The bob is pulled aside until it is in position A. It now has some uphill energy.

a. Where did this uphill energy come from?

b. The bob is released and the pendulum swings back. When it gets to position B, what has happened to the uphill energy?

c. Where will the pendulum swing to after passing B? What happens to the energy?

d. After some time, the pendulum stops swinging, coming to rest with the bob in position B. Where has all the energy gone?

6. A small boy makes a catapult with a piece of elastic. He stretches it and then fires a piece of folded paper across the room so that it hits another boy. Write a short account of this explaining the energy changes at each stage. Your account should mention each of the following: motion energy, strain energy, chemical energy, heat energy.

Other forms of energy

We know that a loud bang can cause our windows to rattle and this tells us that sound is another form of energy. When a brick falls on the floor, most of the motion energy of the brick turns into heat energy, but some will become *sound energy*. Energy is transferred to sound energy by a pop group: they can certainly rattle the windows!

Alpha particle tracks

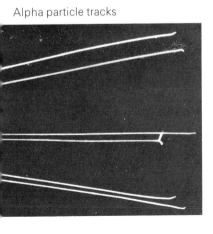

Another form of energy which you will have heard about is *atomic energy*, or *nuclear energy* as it should really be called. There is energy stored up in the nucleus at the centre of atoms. The first photograph shows tracks left by particles 'kicked out' of nuclei of radioactive radium atoms: their motion energy comes from the nucleus. The second photograph, on the next page, shows the sudden release of great quantities of nuclear energy in the explosion of the atomic bomb (really a nuclear bomb). The third photograph shows a nuclear power station in which nuclear energy is released in a controlled reaction: this process may be very important to us in the future as a source of energy. The fourth photograph snows the Sun and it is the release of nuclear energy within the Sun which enables it to send out so much energy to the Earth.

Nuclear power station at Wylfa,
   Anglesey
Atomic bomb
Activity on the sun's surface

But how does the energy from the Sun reach us? The energy clearly travels through the empty space between the Sun and the Earth. There is therefore another kind of energy which we will call *radiation energy*. You have certainly felt this radiation reaching you whenever you have been sun-bathing.

It is this radiation energy which gives us warmth, heating the land and the seas; it is this energy which is absorbed by plants enabling them to grow; without it, we would have no food. The stored chemical energy we have already talked about has in fact all come from the Sun.

It is radiation energy from the Sun which gives energy to the water in the sea, causing evaporation. The water rises and forms clouds, and then the water falls as rain, forming lakes and rivers. In other words the radiation energy from the Sun has given uphill energy to the water.

One form of radiation energy is *light energy*. How do we know that light is a form of energy? It will certainly do jobs of work: it will affect a photographic film so that we can take a picture with a camera, it will cause coloured material to fade, and later you will see in Experiment 13.1l (page 167) light energy turning into electrical energy in an exposure meter.

Homework assignments

1. Read the topic book *Energy*, pages 6–8, and write a short essay on where the energy available in food comes from.

2. Read *Energy*, pages 15–17, to find out where the stored energy in coal and oil comes from. Why are these often called 'fossil fuels'?

3. Read *Energy*, pages 8–12, and write a short essay on the effect of radiation energy from the Sun on the weather.

Bicycle dynamo

Mounted dynamo

Producing electrical energy

The dynamo or generator

You have probably seen a bicycle dynamo like the one shown on the left. No electrical energy comes from it when at rest. But when the bicycle is moving the bicycle wheel turns the dynamo wheel and that causes a current to flow which lights the lamps of the bicycle. The energy flow diagram is as follows.

| food | CHEMICAL ENERGY | body | MOTION ENERGY | bicycle | MOTION ENERGY | dynamo | ELECTRICAL ENERGY | lamp | → LIGHT ENERGY → HEAT ENERGY |

In the second photograph above, a similar dynamo has been mounted so that the lamp lights when the handle is turned.

Turbo-generator in Ratcliffe power
station

Electrical generators in power stations work in a similar
way (although their design is obviously different from the
bicycle dynamo). The generator has to be turned in order to
produce electrical energy. The problem for the engineer is to
find some way of producing the turning effect.

The water turbine

In the mill above, the water turns a wheel which drives the
grinding mechanism inside the mill.

This is the principle of the water turbine. A jet of water hits the blades inside the turbine, causing them to turn, and the turbine then turns the generator.

| dam | $\xrightarrow{\text{UPHILL ENERGY}}$ | water | $\xrightarrow{\text{MOTION ENERGY}}$ | turbine | $\xrightarrow{\text{MOTION ENERGY}}$ | generator | $\xrightarrow{\text{ELECTRICAL ENERGY}}$ | lamp | → HEAT ENERGY → LIGHT ENERGY |

The greater the motion energy of the water the better, and that is why hydroelectric power stations are found where there are great falls of water (Niagara Falls and Victoria Falls, for example). Sometimes large dams are built to create artificial lakes, and hydroelectric power stations are put at the foot of such dams. The height of the reservoir provides sufficient uphill energy to drive the turbines in the power station.

Dams and power stations in France (left) and Switzerland (right)

The steam turbine

If a fall of water is not available, a different type of turbine is used in which the blades are turned by steam. The steam is produced by heating water either by burning oil, gas or coal or by using the heat generated in a nuclear reactor. The energy flow diagram for a fuel burning power station is:

Experiment 13.1   Looking at energy transfer

a. Lighting a match

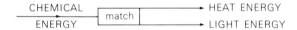

b. Using a Bunsen burner

c. Hammering a thin strip of lead

d. Using a battery to light a lamp

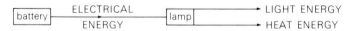

e. Using a model steam engine to raise a load

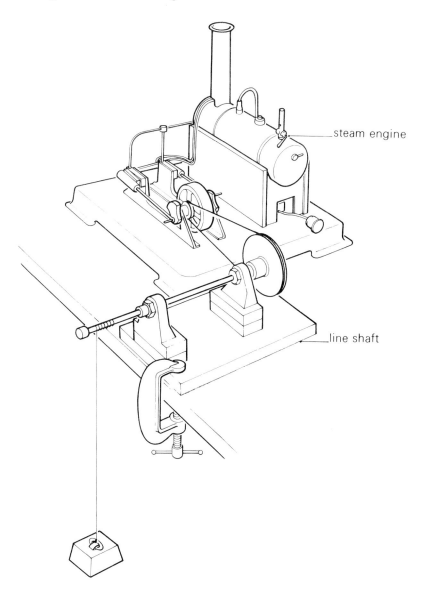

steam engine

line shaft

f. Using a model steam engine to drive a dynamo to light a lamp

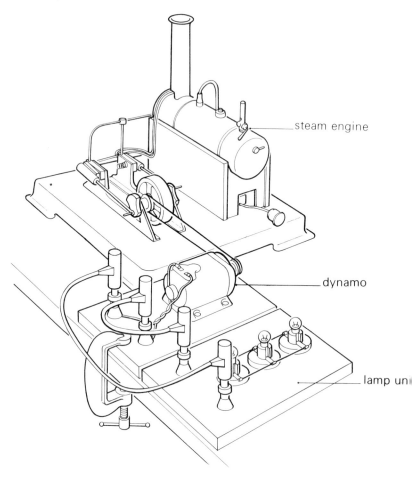

steam engine

dynamo

lamp un

g. Using a battery to drive a motor, which drives a dynamo which lights a lamp

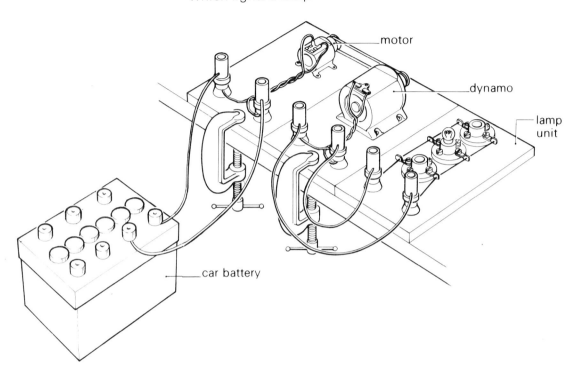

motor

dynamo

lamp unit

car battery

h. Using a battery to drive a motor to lift a load

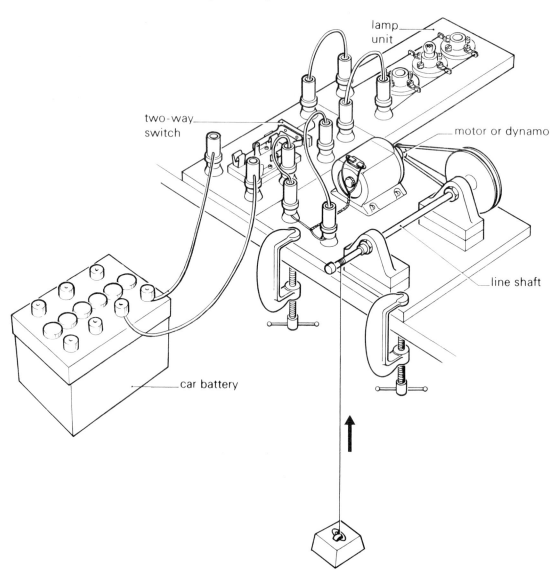

When the load has been lifted, disconnect the battery by changing the switch. Then let the falling load drive the motor as a dynamo, which lights a lamp.

i. A storage cell

A form of storage cell can be made with two lead plates in dilute sulphuric acid. A battery is connected across the lead plates for a few minutes. The storage cell is disconnected and then connected across a lamp which glows for a short time. Electrical energy from the battery is transferred to chemical energy in the storage cell, and then this becomes electrical energy again which lights the lamp.

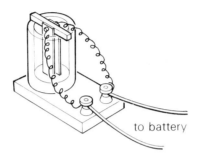

to battery

j. Use a jet of water to drive a turbine, which in turn drives a dynamo, which lights a lamp

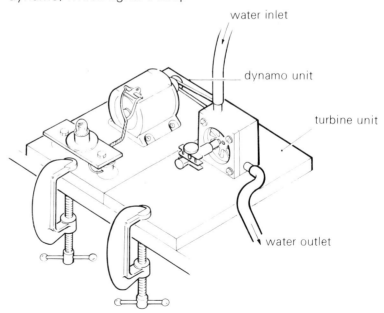

water inlet

dynamo unit

turbine unit

water outlet

k. Use the turbine in reverse as a pump

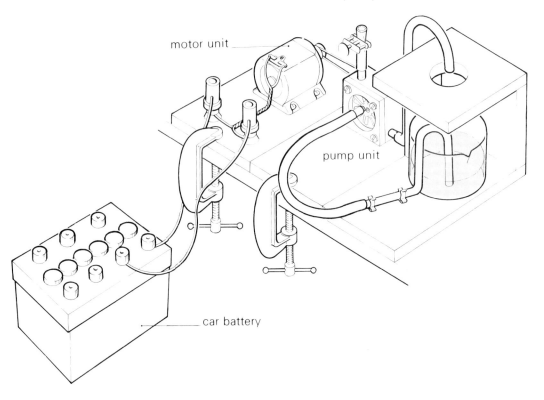

motor unit

pump unit

car battery

In this experiment, a battery drives a motor, which turns the turbine unit as a pump so that water is taken from a lower level to a higher one. The energy flow diagram is as follows:

battery	ELECTRICAL ENERGY	motor	MOTION ENERGY	pump	MOTION ENERGY	water	UPHILL ENERGY

l. A coupled pendulum

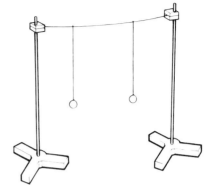

This experiment works well if the stands are rigidly clamped to the bench and the lengths of the two pendulums are exactly the same. Set one of the pendulums swinging and watch the energy being transferred to the other and then back again to the first.

m. Photographic exposure meter

When light falls on the photocell in the exposure meter, it causes a current to flow and the pointer moves. Light energy is converted into electrical energy and then into spring energy.

light falls on photocell

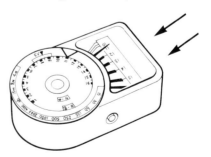

n. Using a motor to drive a flywheel

Energy from the battery is transferred to the motor, which drives the flywheel. If the switch is changed, the flywheel will drive the motor as a generator, which will light a lamp.

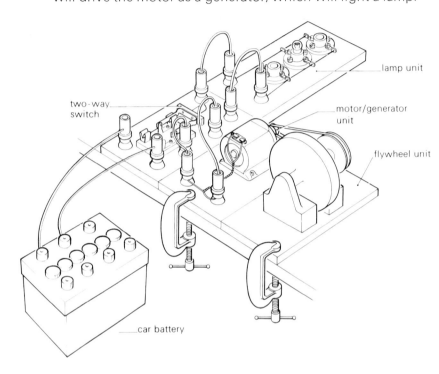

lamp unit

two-way switch

motor/generator unit

flywheel unit

car battery

Questions for homework

1. Draw a flow diagram for the energy transfers in the following experiments.

a. A model steam engine raising a load (Experiment e above).

b. A model steam engine driving a dynamo, which lights a lamp (Experiment f above).

c. A battery which drives a motor, which drives a dynamo, which lights a lamp (Experiment g above).

d. A water supply which produces a jet of water, which drives a turbine, which turns a dynamo, which lights a lamp (Experiment j on page 165).

e. A motor which drives a flywheel, which later drives a dynamo, which lights a lamp (Experiment n above).

2. It is said that when the scientist Joule was on his honeymoon in Switzerland he spent time trying to show that the temperature of the water at the bottom of a waterfall was higher than the temperature at the top. Do you think it might be true that there is a temperature rise? Give a reason.

3. Chemical energy from the petrol is necessary to start moving a car from rest.

a. Some of the chemical energy is transformed to motion energy of the car, but not all of it. Suggest what else the chemical energy gets turned into.

b. When the car is travelling at a steady speed, it still needs a supply of petrol to keep it going even though the motion energy stays the same. Can you suggest why this is necessary?

4. A motor car starts from rest. It then climbs a steep hill and comes to rest on top. Unfortunately, the driver leaves the brakes off and the car starts to move down the hill, going faster and faster, until it hits a tree at the bottom of the hill. Describe the energy changes that occur throughout both parts of this sad story, or draw an energy flow diagram for each part.

5a. Coal is burned at a power station to make steam. This steam is used to drive a steam turbine which is coupled to an electric generator. Describe the energy transformations which take place.

b. A company uses electric power to drive an electric motor coupled to a pump, which pumps water from a deep well to a reservoir on a hilltop. Describe the energy transformations in this process.

c. How could the store of water in the reservoir be used to produce electrical energy?

d. How could the energy stored in the reservoir be increased?

6. Write an essay on energy, mentioning the different forms it may take. Give examples of those you mention.

match stick

drawing pin

candle wax

7. A toy 'tank' can be made from a cotton reel, an elastic band, a pencil, a match stick, a drawing pin and a thin slice cut from a wax candle. Such a tank is illustrated on the left.

If you wind up the elastic band using the pencil and put the tank on the floor it will creep along.

a. Where does the tank obtain the energy to move?

b. What is the purpose of the match stick and the drawing pin?

c. Is the pencil necessary? Would the tank work as well if the pencil were replaced by another matchstick?

d. Why do you think the thin slice of candle is used?

e. Do you think that such a tank, placed on a very smooth surface which is slightly tilted, would travel the same distance uphill as it would downhill before coming to rest? Give a reason for your answer.

Homework assignments

1. Read the topic book *Energy*, pages 38–40, and prepare a short talk on hydroelectric energy.

2. Read *Energy*, pages 33–38, and prepare a short talk on solar furnaces and solar cells.

3. A possible source of energy is energy from the tides. This has been tried on the Rance River in France. Read about this topic in *Energy*, pages 41–42. Study a map of Great Britain to see where there are sites where a dam might be built. Write a short essay about the advantages of using energy from the tides, and the requirements of any site you chose.

4. Electricity is produced in hydroelectric power stations using the uphill energy stored in water at a high level. The electricity produced is fed into the national power line system. Many hydroelectric power stations use some of the electricity they produce at certain times of the day to pump water back from the lower level to the higher level. This may seem a strange idea. What is the reason for it?

5. Find out what you can about one of the following and write some notes on how it works: Newcomen's steam engine, a petrol engine, Watt's steam engine, a diesel engine.

6. Look at the energy consumption graph for the United Kingdom shown on page 46 of *Energy*. What is happening at the present time and what might happen in the future?

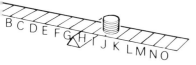

Chapter 14 **Revision tests C**

Revision test C1

1. The diagram shows six pennies of equal mass placed on a uniform piece of wood at the line J, two units to the right of the fulcrum at H.
a. Where would you place two pennies to balance the lever? (Assume that all the pennies in this question are similar and therefore have the same mass.)
b. Suggest another way of balancing the lever using three equal pennies.
c. With the six pennies still at J and one penny at A, how could a balance be obtained?

2. When smoke particles suspended in air are strongly illuminated and viewed through a microscope, they show small, random, irregular movements.
a. What is the explanation of the movement?
b. What does this experiment tell us about the nature of the air?
c. How do you know that the movements are not caused by vibration of the apparatus?
d. If the air is warmed, would you expect to see any general change in the irregular movements? If there is a change, why does it occur?

3. The diagram shows a simple barometer set up in the laboratory. Copy the diagram and mark points with the following letters:
a. an E for a point in the liquid at which the pressure is equal to that of the atmosphere,
b. an L for a point in the liquid at which the pressure is less than that of the atmosphere,
c. an M for a point in the liquid at which the pressure is more than that of the atmosphere.

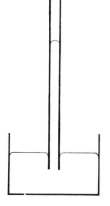

4. In order to estimate the diameter of a very small ball-bearing, a large number of ball-bearings was poured into a box lid 20 cm long and 15 cm wide so that as many as possible were used to form a single layer of balls on the bottom of the lid.

a. What is the area occupied by the layer of ball-bearings?

  The ball-bearings were then carefully poured into a graduated cylinder containing 10 cm³ of water. The level of the water rose to the 40 cm³ mark.

b. What is the volume occupied by the ball-bearings?

c. From these two results, it is possible to make a very rough estimate of the diameter of a ball-bearing. What is your estimate of it?

d. Do you consider that this is a good method of estimating the diameter? Give your reasons.

5. Radiation from the Sun falls on the sea, causing some of it to evaporate. Clouds form and rain falls, some of which goes into a reservoir high in the mountains. Water flows down from the reservoir and drives turbines which turn generators to produce electricity which is used to provide heat and light in your home. Give a list of the energy transfers occurring throughout this, in the order in which they occur.

Revision test C2

1. Different masses M are attached one after another to a rubber band. The extension is measured for each mass and a graph is plotted of the readings.

a. What would be the extension if a mass of only $\frac{1}{2}$ kg were attached to the rubber band?

b. Two similar rubber bands are joined together and a mass of 2 kg is attached (see Figure iii). What would the total extension be?

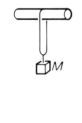

i

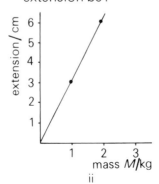

ii

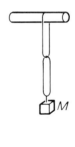

iii

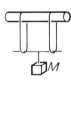

iv

c. What would the extension be if the rubber bands are put side by side as shown in Figure iv and the same mass of 2 kg is attached?

2. A bottle has a label on the side saying 'vacuum'. Various suggestions are made for finding out if there is a vacuum inside. Write a few lines commenting on each suggestion, saying what would happen in each case. Mention which method you think would be the best.

a. 'Open the bottle and put a lighted match inside and see if the match goes out.'

b. 'Measure the mass of the bottle on a balance, then open the bottle and find its mass again to see if there is any difference.'

c. 'Heat the bottle in boiling water and see if the pressure increases so that the top blows out.'

d. 'Immerse the bottle completely in a bucket of water and then open it.'

3 a. Explain briefly what is the difference between *force* and *pressure*.

b. If you were condemned to lie on a bed of nails, would you choose one with a very large number of nails or a very small number? The nails are all the same size. Explain your choice, assuming that you are no better than the rest of us at putting up with pain.

c. If a person chooses 4 000 nails, each with a square tip of side $\frac{1}{40}$ cm for his first attempt, and at the second attempt chooses 1 000 nails with square tips of side of $\frac{1}{10}$ cm, which would be the more painful experience? Give the reason for your answer.

4. 'Molecules are very small'. This opinion is supported by the Oil Film experiment. Describe what is done in the experiment.

Explain how it supports the idea that molecules are small.

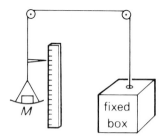

5. A light string, connected to a scale pan which has a mass of 10 g, passes over two pulleys and then through a hole in the top of a fixed box.

A marker shows the different positions when a mass m is added to the pan. The following readings are obtained.

load m/g	0	5	10	15	20	30
scale reading/cm	2	3	4	5	6	8

a. What might be attached to the string inside the box? Give a reason for your answer.
b. What scale reading would correspond to $m = 25$ g?
c. What scale reading would correspond to $m = 12$ g?
d. When asked what the scale reading would be for a load of 60 g, one boy says the answer is 14 cm, another says 16 cm and a third says that it is impossible to tell. With which of these do you agree? Give your reason.
e. If the scale pan is replaced by one with mass 20 g, what will the scale reading be when a load of 10 g is put in it?

Revision test C3

1. Three cubes are placed on some soft Plasticine. Cube A has a mass of 72.0 g and side 2 cm. Cube B has a mass of 72.9 g and has a side of 3 cm. Cube C has a mass of 64.0 g and a side of 4 cm. Which of the cubes will sink furthest into the Plasticine? (Show any working you do.)

If the density of aluminium is 2.7 g/cm³, which of the cubes is aluminium? Give your reasons.

2. A see-saw is made of a plank 4 metres long supported at its mid-point. A man of mass 80 kg sits so that his weight acts exactly over one end. A boy of mass 50 kg sits similarly at the other end and beneath him is suspended a very light polythene bucket. What mass of water will have to be put into the bucket to make the plank exactly horizontal?

If the bucket is now moved 1 metre nearer to the pivot (the man and boy remaining at the two ends), what mass of water will now be necessary? Explain how you arrive at your result.

3. The figures represent four barometers. The level of the mercury is shown in figure i.

There is a vacuum above the mercury, and the mercury levels in all the four dishes are the same. In Figure ii the tube is twice as wide as that in Figure i. In Figure iii the tube is half as wide as in Figure i. In Figure iv the tube is the same width as in Figure i but tilted.

Copy the figures and carefully mark the level of the mercury in all the tubes.

What difference, if any, would it make in Figure i if a little air got into the top of the tube?

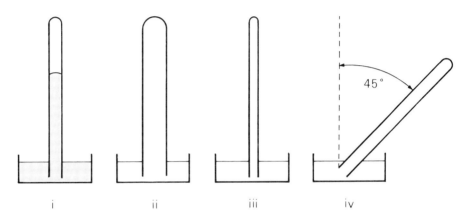

i   ii   iii   iv

4a. Explain briefly why the reservoir of a public water supply is situated on high ground.
  b. Two taps, one upstairs and one downstairs, are both supplied from the same tank in the roof. When the taps are turned on by the same amount, the water runs out of the upstairs tap more slowly than from the other. Explain this.
  c. A boy of mass 60 kg is walking on snow with his father whose mass is 90 kg. The soles of the boy's shoes have an area of 30 cm² each and those of his father 40 cm² each. Who will sink further into the snow? Give reasons for your answer.
  d. Explain why the wall of a dam is thicker at the bottom than at the top.

5. A rubber ball is resting near the edge of a table. You push it forward off the table and it falls to the ground, bouncing several times before coming to rest. Copy the diagram and mark on it the path the ball will take.

Make a list of the energy changes which occur. What has happened to the energy when the ball comes to rest?

Revision test C4

1. Explain the following.
a. The pressure of the atmosphere decreases with height.
b. A closed treacle tin contains nothing but air, but the lid blows off if the tin is heated.
c. Astronauts working on the Moon have space suits which appear puffed up, although the suits do not appear puffed up when they are worn on Earth.
d. A piece of cork and a lump of lead of the same volume are held under water. When they are released, one rises and the other sinks.

2. Movement can be seen when smoke particles in a draught-free cell full of air are suitably illuminated and viewed through a microscope.
a. Describe briefly the type of movement which can be seen.
b. What is the explanation of the movement?
c. A speck may seem bright for a few seconds, then it may become blurred or even disappear. Why is this?
d. Some of the particles move less rapidly than others. Are these the larger particles or the smaller particles? Give a reason.

3a. 'Hickory dickory dock, the mouse ran up the clock . . .' What energy change took place when this happened?
b. 'The clock struck one . . .' Where did the clock obtain the energy for this?
c. 'The mouse ran down . . .' What energy changes took place when the mouse did this?

d. 'Since the mouse ends up where he started from he has neither gained nor lost energy.' Do you agree with this statement? Give reasons for your answer.

4a. You are given a barometer tube (closed at one end), a small glass beaker and mercury. Describe how you would set up your own barometer and draw a diagram of it.

b. Suggest a reason why a barometer set up in this way might give an inaccurate measurement of atmospheric pressure.

c. Would the reading be high or low?

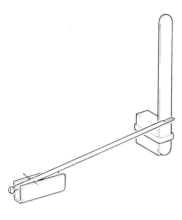

5. The drawing shows a simple microbalance made with a drinking straw. You also have a packet of 100 sheets of graph paper and you know the packet has a mass of 400 g. Describe carefully how you would use the apparatus to find the mass of a hair. Explain how you would calculate the answer.

Descriptive revision test C5

1. Describe the oil film experiment for estimating the size of an oil molecule.

2. What apparatus do you need to show Brownian motion? Explain how it would be set up and what you would see.

3. What is meant by *diffusion*? Describe an experiment which shows diffusion between gases and another to show diffusion between liquids.

4. How can you demonstrate the change in volume when a liquid turns to a gas?

5. Describe an experiment to show motion energy turning into heat energy, and another showing heat energy turning into motion energy.

Chapter 15 **Electric circuits**

Electricity plays a very large part in the lives of us all, so much so that we tend to take it very much for granted. We are very used to switching on the electric light, an electric heater, a record player, a radio, or a spin dryer. Cars would not work without electricity, and aeroplanes would not fly. The telephone would not exist without it and there would be no television. In our homes we would once more have to use oil lamps or gas lighting. We owe a great debt to those scientists whose investigations made all modern electrical engineering possible.

There is another reason for studying electricity. It has been found that the study of matter is closely related to the study of electricity. If we are to learn more about atoms, it is essential to have a knowledge and understanding of electricity.

In this chapter we will be looking at various electric circuits, and in the next chapter at some of the effects of an electric current. A lot of experiments will be discussed and of course it is most important that you do these experiments yourself. It is through them that you will come to understand electric circuits.

Experiment 15.1   Using a cell and a lamp

a. Fix the cell in the circuit board. Put the lamp into its holder and fix this between two of the pegs on the board. Use two leads to make the lamp light.
b. Find out what happens if you put the cell in its holder the other way round. Does it make any difference?
c. Does it make any difference if you put the lamp holder the other way round?
d. Does it make any difference to the brightness of the lamp where you put the holder on the circuit board? Does the shape of the circuit matter? Does it make any difference whether the leads are long or short?

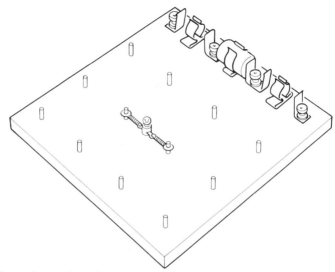

The electric circuit

These experiments show that a complete conducting path is necessary for the electricity to flow. We refer to this complete conducting path as an electric circuit. The lamp does not light if there are any gaps in the circuit. It is usual to speak about an electric current flowing round an electric circuit, though what exactly we mean by a current will not be clear until later.

Experiment 15.2   Using several lamps and several cells

Put two lamps in line and connect the cell to the ends as shown in the drawing on the next page. How does the brightness compare with the brightness when the cell was connected to only one lamp?

The word battery. In everyday use, the word *battery* is often used instead of the word *cell* used above. Strictly speaking, a battery consists of a number of cells, in the same way that a battery of guns consists of a number of guns. We shall use the scientific word *cell* throughout this book and only use the word battery when we mean a number of cells.

Put three lamps in line and connect one cell across them (in other words, connect one side of the cell to one end of

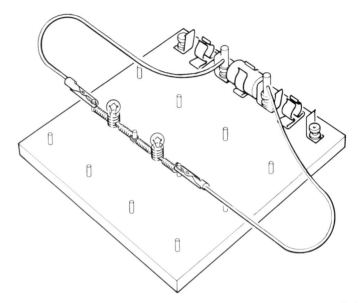

the line and the other side of the cell to the other end of the line). What happens to the brightness? (If by any chance you cannot see the lamp glowing, shield it with your hand and look closely for a faint glow.)

Now try two cells and two lamps. How does the brightness compare with the brightness when one cell was connected to only one lamp?

Leave one of the lamps alone, but turn the other round. Does it make any difference to the brightness?

Leave one of the cells alone, but turn the other one round. Does it make any difference to the brightness?

Fix three cells on the circuit board, all of them the same way round. Put three lamps in line and connect the cells across them. What happens to the brightness?

Now connect the three cells across two of the lamps. What happens to the third lamp? What happens to the brightness of the other two?

Then connect the three cells across one of the lamps. What happens to this lamp?

Normal brightness of a lamp

Connect a single cell across one lamp. Notice the brightness of the lamp. In future we will call that *normal brightness*.

When two cells are across one lamp, the brightness is

greater than normal brightness. And when three cells were across one lamp, the brightness was very much greater than normal brightness; in fact you had a miniature photoflood or it may have been so bright that the lamp burnt out.

When one cell was put across two lamps, they glowed only dimly. The brightness was less than normal brightness. With one cell across three lamps, the brightness was even less and the lamps glowed very dimly indeed.

One lamp across one cell glowed with normal brightness. But so did two lamps across two cells. What happens with three cells across three lamps? Try it and see. They all glow with normal brightness.

The answer to that last question is the answer the author of this book, and probably your teacher, would expect you to give and it is really the 'correct' answer. But it may happen that someone in your class will put three cells across three lamps and they may not all glow with normal brightness, in fact one may be slightly brighter, one might be slightly less bright. Is this 'wrong'? Is there some 'fault'? Certainly not, he has done the experiment and that is what happened. There are often slight variations between one lamp and another, even though they are made by the same manufacturer. Lamps from different manufacturers can differ quite a lot even though they are supposed to be the same. Cells can also differ from one another (especially when one has been used a lot and another has not). This is the real, everyday, practical world in which we live. Your teacher will probably do his best to see that your lamps are as nearly the same as possible to make matters easier, but we must take a sensible view when there are slight variations. On the whole it is true to say that three lamps across three cells will glow with the same normal brightness of one lamp across one cell.

Experiments 15.3   Lamps in series and parallel

When lamps are arranged in line, as in the left hand drawing below, they are said to be *in series*. Another arrangement is shown in the right hand drawing in which the lamps are placed side by side. These lamps are *in parallel*.

Put two lamps in series on your circuit board with two cells across the ends. The lamps should glow with normal brightness.

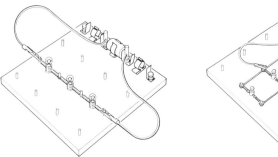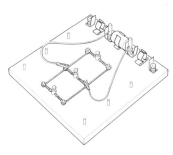

Now connect the two lamps in series with only one cell across the ends. What happens to the brightness?

Disconnect the two lamps and reconnect them in parallel. Put one cell across the ends. What happens to the brightness this time?

What do you think would be the difference in brightness if you put one cell across three lamps in series and then put it across three lamps in parallel? When you have decided, try the arrangements on the circuit board and see if you were right.

Circuit diagrams

To draw a picture of three lamps in parallel across two cells would be an awkward business if it had to be drawn like the drawings above. For this reason scientists adopt a special way to record electric circuits. There are certain special signs or symbols which are usually used in these diagrams.

The symbol for a cell is two parallel lines as shown on the left, one longer and thinner than the other. We have already seen that it matters which way round a cell is used in a circuit. You will notice that some cells are labelled +and —. With your cells, the central 'button' is the positive terminal

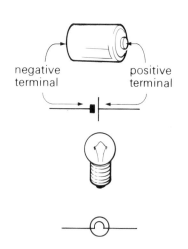

negative   positive
terminal   terminal

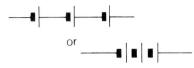

or

and the metal base at the other end is the negative terminal. In the symbol for the cell, the long line represents the positive terminal, the short line the negative one. The second drawing shows the standard symbol for a lamp.

On your circuit boards you sometimes had three cells in series (connected + to − in a line). This can be shown on a circuit diagram in either of the ways shown on the left.

Several cells in series might be shown as:

You found in your experiments above that it was necessary to have a complete circuit in order to get a lamp to light. The complete circuit must be shown in your diagram with lines. Where lines meet it is usual to mark the junction with a dot as shown on the left.

The circuit for one lamp and one cell would be drawn like this:

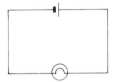

And the circuit mentioned above with three lamps in parallel across two cells might be drawn in either of these ways (they are different ways of drawing the same thing):

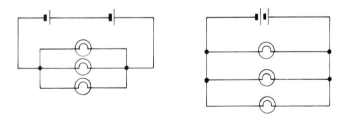

Questions on circuit diagrams
1. Draw a circuit diagram for each of the following:
a. One cell across two lamps in series.
b. One cell across two lamps in parallel.
c. Three cells across three lamps in series.

2. What is wrong with these circuit diagrams if you want all the lamps to light? Copy the diagrams and put them right.

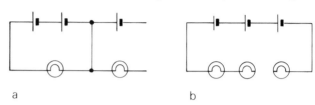

a   b

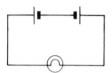

3. Look at this circuit carefully. Do you think the lamp will glow brighter than normal, with normal brightness, dimly or not at all? Give a reason for your answer.

4.

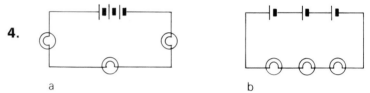

a   b

These two circuit diagrams are drawn differently. Are they electrically the same or different?

5.

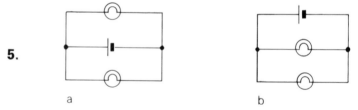

a   b

These two circuit diagrams are drawn differently. Are they electrically the same or different? Give a reason.

6.

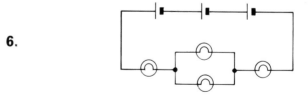

Set up this circuit on your circuit board. Which lamps glow brighter than normal and which less bright than normal? Explain what you see.

Experiments 15.4 Circuits with switches

You have already found that you must have a complete circuit; there must not be a gap in the circuit if the lamp is to light. This fact makes it possible for you to use a very simple switch in your circuits, as shown in the drawings below. With the switch in the open position there is a gap in the circuit; but when the switch is pressed, the gap is closed and the circuit is completed.

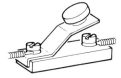

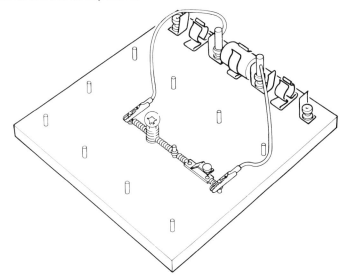

With your circuit board, connect a cell, a lamp and a switch in series so that you can switch the lamp on or off.

Now put two cells, two lamps and a switch in series so that you can switch both lamps on or off at the same time.

Put two lamps in parallel across a cell, so that they both glow with normal brightness.

Now add two switches to the circuit so that one switch operates one of the lamps and the other operates the other lamp.

When you have solved it, draw a circuit diagram of your arrangement. Use this symbol for a switch.

Experiments 15.5 Conductors and insulators

You know that an air gap prevents a current from flowing in a circuit. We say that air is an *insulator*. A strip of metal and the leads you have been using allow a current to flow: these are therefore called *conductors*.

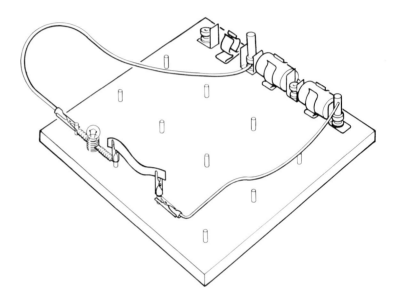

Set up the circuit shown in the drawing. The crocodile clips are a convenient way of fixing various objects in the circuit to find out if they are conductors or insulators. If they are conductors, the lamp will light; if they are insulators, it will not.

Collect as many different objects as you can and test each of them. Make a list showing which are conductors and which are insulators.

Your investigation might include the following, though it is hoped you will try other things as well.

strip of paper	a wooden pencil	tin can
piece of wood	pencil lead	protractor
strip of copper	aluminium foil	plastic cup
thread of nylon	comb	saucer
piece of material	handkerchief	hair
piece of lead	knitting needle	expanded polystyrene

Experiments 15.6   Resistance wire

For this experiment you are provided with some special wire: eureka wire is the name for it, but there is no need to remember that name.

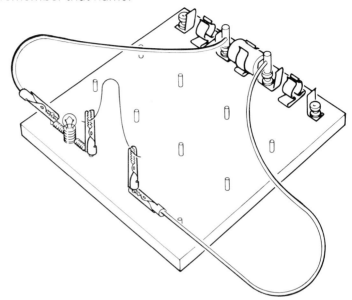

Use the same apparatus as in Experiment 15.5, but only one cell is necessary for this experiment. Put a very short length of the wire between the crocodile clips. Does the lamp light or not? Is the wire a conductor or an insulator?

Then try a longer piece of eureka wire between the crocodile clips. Is there any change in the brightness of the lamp?

Now try a very much longer piece of the wire. What happens to the lamp now? If your piece of wire is very long, see it does not touch part of itself. What happens if it does?

The dimmer

The previous experiment has shown that when a short length of eureka wire is used, it seems to be a good conductor and the lamp glows brightly. But as a longer and longer piece is used, it is harder for current to flow and the lamp gets less bright. It appears to discourage the current and for that reason such wire is called *resistance wire*.

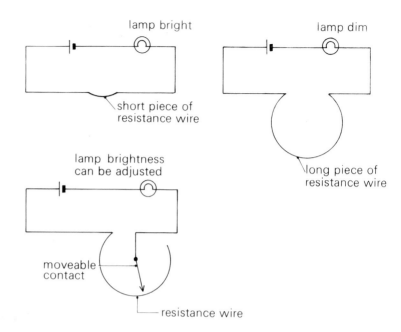

lamp bright

short piece of resistance wire

lamp dim

long piece of resistance wire

lamp brightness can be adjusted

moveable contact

resistance wire

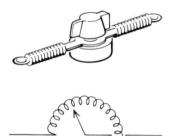

With a short length of resistance wire in the circuit, the lamp was bright. With a long piece the lamp was dim. This suggests that such a length of wire included in a circuit would make a very good *dimmer* for changing the brightness of lights. But a long piece of wire like that would be very inconvenient. So manufacturers wind it up into a convenient coil with a control knob on top which varies the length of wire through which the current flows.

A piece of wire or other substance which offers some resistance to a current is called a *resistor*. A dimmer is sometimes called a *variable resistor* and another name sometimes used for it is a *rheostat*. In a circuit diagram, the symbol we will use for a resistor is

and the following symbol represents a variable resistor:

Experiment 15.7 The dimmer

Set up a circuit board with two cells across one lamp so that it glows more brightly than normal.

Insert a dimmer in the circuit so that you can control the brightness of the lamp. Does it make any difference which side of the lamp you put the dimmer?

Now set up the circuit on the left on your circuit board with two lamps in parallel.

Insert two dimmers in the circuit so that you can control the brightness of the two lamps independently of each other.

Experiment 15.8 The rectifier

Set up a circuit board with two cells, a lamp and two crocodile clips as shown. Complete the circuit by connecting the rectifier between the two crocodile clips. What happens to the lamp?

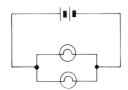

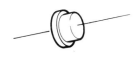

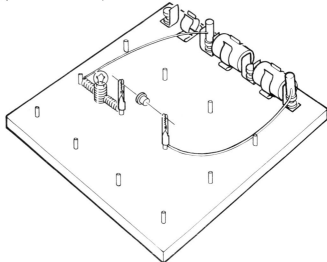

Now turn the rectifier round. What happens to the lamp this time?

This little device has a very high resistance one way round (it is virtually an insulator) and a low resistance the other way round. In effect it allows current to travel one way through it, but not the other way. In circuit diagrams, rectifiers are usually shown by the symbol on the left.

The current flows in the direction of the arrowhead and it will not flow the other way round.

If you have time, you might like to try the following circuits on your circuit board.

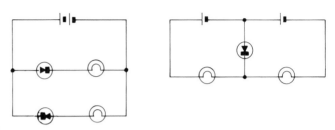

Each time, when you have set it up, turn the cells round to see what difference it makes. Can you explain what happens?

Questions for homework or class discussion

1. When a lamp is lit by a single cell, we shall say that it has *normal brightness*.

All the cells in the circuits below are similar, and all the lamps are similar. State whether each of the lamps will be extra bright, at normal brightness, dim or out.

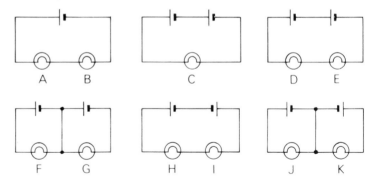

2. When a lamp is lit by a single cell, let us say that it is lit with normal brightness. You are given three similar lamps and one cell. Draw the circuit diagrams that would give each of the following results:
a. all three lamps glow dimly with the same brightness,
b. all three lamps glow with normal brightness,

c. two lamps glow equally dimly and one with normal brightness.

What will happen if one lamp is unscrewed from its holder in b? In c, what will happen if one of the dimly lit lamps is unscrewed?

3. The circuit below shows three cells lighting three lamps.

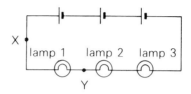

The points X and Y are joined by a piece of copper wire (a good conductor). What will happen to the brightness of lamp 1? What will happen to the brightness of lamps 2 and 3?

4. What would you expect to happen to the lamp in the circuit on the left:

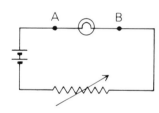

a. if the resistance of the variable resistor is reduced?
b. if a piece of copper wire (a good conductor) is connected between the points A and B?
c. if a second, similar lamp is connected to the points A and B in parallel with the first lamp?
d. if one of the cells is turned round?

5. In each of the circuits, Figures i, ii and iii, describe what changes will happen to each lamp when the switches are closed (they are drawn in the open position).

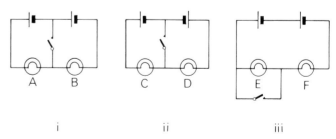

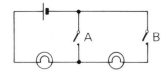

6. This circuit shows one cell, two similar lamps and two switches A and B (shown in the open position). When the cell is connected across one lamp it glows with *normal brightness*. Copy the following table and describe whether on each occasion the lamp will be bright, dim or out.

Switch position	Lamp 1	Lamp 2
A open, B open		
A open, B closed		
A closed, B open		
A closed, B closed		

7a. Describe the effect of opening and closing the switches in each of these circuits.

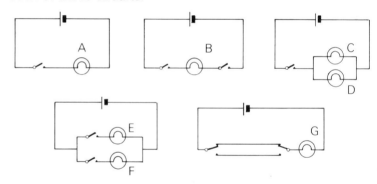

b. Would the second circuit be suitable for turning a light on from both the front door and the back door of a house? What would be the disadvantage of this circuit? Which circuit would be best for this purpose?

Chapter 16 **Electrical Measurements**

In Chapter 7 we discussed magnetic forces, how one magnet exerts a force on another. You are familiar with small compasses which contain a small magnet (or 'needle' as it is sometimes called). One end is always turned towards the North and the other to the South. In the experiment that follows you will find that a current passing through a coil of wire can influence a compass needle; the current in the coil has a magnetic effect.

Experiment 16.1   Magnetic effect of a current
Set up a circuit board with two cells and one lamp as shown. Wind a coil of wire loosely round a pencil and then fix the bared ends of the coil to the crocodile clips. Put a small compass next to the coil. It will point North and South, and the coil should be arranged so that it is East or West of it.

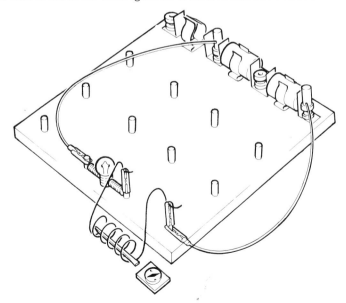

Switch the current on and off. Does the coil have any effect on the compass needle?

Take an iron nail. Put it in the coil. Switch on the current through the coil. Is the effect on the compass needle now much greater?

Experiment 16.2 Measuring current with a current balance

So far we have measured the strength of an electric current by the brightness of a lamp, in the circuit. When the lamp is dim, we say the current is small; when it is very bright, we say the current is large. This is not a very precise way to measure current. We can use the magnetic effect of a current to make a current balance which will measure current for us.

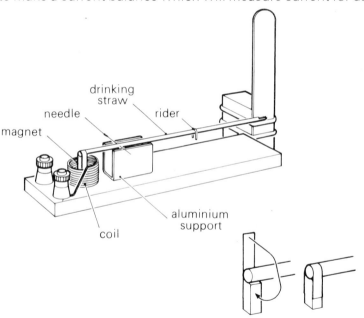

Use Sellotape to fix a small magnet to one end of a drinking straw. Put the needle through the straw and balance the straw on the metal support so that the lower end of the magnet is near the centre of the circular coil. Arrange a small rider (made from, say, 2 cm of bare copper wire) on the straw so that the straw is horizontal. Make a reference mark on the wooden strip, attached to the wooden block.

When a current is passed through the coil round the magnet, the magnetic effect causes the small magnet to be attracted or repelled. The straw goes up or down by an amount depending on the strength of the current. The rider can then be moved to bring the end of the straw back to the reference mark. The distance it is moved is a measure of the strength of the current.

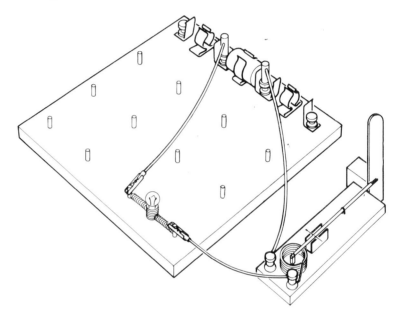

Set up a circuit board with one cell and one lamp. The lamp glows with normal brightness and we will call the current that flows 'one lamp's worth of current'. Insert the current balance in the circuit. Move the rider so that the straw comes to the reference mark. The current balance now measures one lamp's worth of current.

Try these circuits. Find out in each circuit whether the current is one lamp's worth, more than one lamp's worth or less than one lamp's worth.

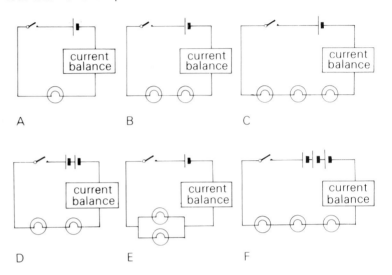

A B C

D E F

Homework assignments

1. Read the topic book *Magnetism* and prepare a short talk on one of the following topics: early history of magnetism, simple magnetic effects, methods of making a magnet, magnetic maps, the Earth's magnetic field.

   Remember that a talk is always more interesting for your audience if you have something to show them. Think carefully about what experiments you might do to make the talk interesting.

2. Read the topic book *Electric currents*, pages 20–25, on the magnetic effects of an electric current. Give a short talk on electromagnets.

3. See if you can obtain an old electric bell or an electric relay. Prepare a short talk on how it works.

Ammeters

As a way of measuring a current, a 'lamp's worth' is not very satisfactory since there are many different kinds of lamp and even lamps which are supposed to be the same may differ from one another. It is more usual to measure current in a unit called an *ampere* (often abbreviated to A). An instrument used for measuring current is an *ammeter* (from the words 'ampere meter').

Mass-produced commercial ammeters are much easier to use than current balances and you will doubtless prefer to use them in future.

Ammeters

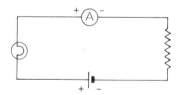

In order to avoid damaging an ammeter, it is important to connect it in a circuit the right way round. One terminal of an ammeter is often coloured red or it may be marked +. This terminal of the meter must be connected to the wire which eventually goes to the + terminal of the battery or cell. The black or − terminal is connected to the wire which eventually goes to the − terminal of the battery or cell. There are lamps or other items in the circuit, but the meter must always be connected this way round.

The symbol used in circuit diagrams for an ammeter is : Ⓐ

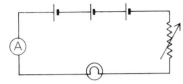

Experiment 16.3   Currents in circuits

a. Set up the circuit shown. The current can be altered by changing the variable resistor. Notice the reading of the ammeter when the lamp is very dim. Adjust the resistance until the lamp is at normal brightness. Then adjust until the lamp is very bright. How much current flows each time?

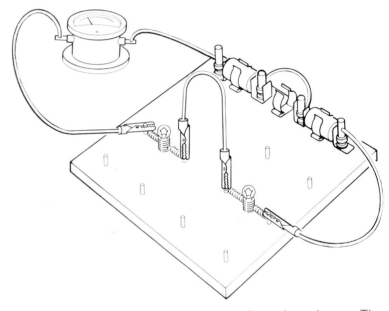

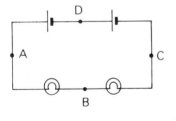

b. Set up the circuit shown, using two cells and two lamps. The drawing above shows the ammeter inserted in the circuit at A. What does the ammeter read?

Put the ammeter into the circuit at B instead of A. Notice the ammeter reading. Then put it at C and again notice the reading.

Finally insert the ammeter at D between the two cells. What is the ammeter reading this time?

c. Set up this circuit and insert the ammeter at each of the points A, B, C, D, E, F.

From Experiment b we would expect the readings at C and D to be the same. Are they? How do the readings at B and E compare? And those at A and F?

How does the reading at A compare with the reading at B and that at C?

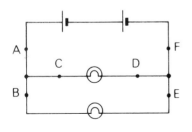

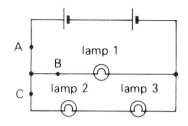

d. Look at this circuit. Will an ammeter at B read more than, the same as or less than an ammeter at C? Then decide how the reading of an ammeter at A will compare with the readings at B and C. Having forecast what will happen, set up the circuit and see if you were right.

Currents in electric circuits

Some important facts about electric currents follow from the experiments which you have done.

In a simple circuit the current is the same all the way round the circuit
It is untrue to say that 'the current gets used up going round the circuit': it is the same current everywhere, even between the cells of the battery. It is the same current through each component in the circuit, including the cells themselves.

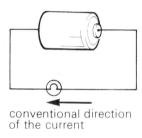

conventional direction
of the current

Currents have direction
In fact, none of our experiments have shown which way the current is flowing, but it is a convention to assume that the current flows through the circuit from the positive to the negative terminal of the cell.

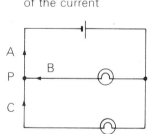

As much current leaves a junction as reaches it
In the circuit on the left, the current reaching the point P is $B + C$. The current leaving the junction is A. It therefore follows that current $A =$ current $B +$ current C. This result is not really very surprising as the following may show.

Suppose two roads P and Q merge into a wide road R, and that they are one way streets. If there is one car travelling along road P every hour and if there are two cars per hour along road Q, how many cars per hour does common sense tell us must be travelling along road R?

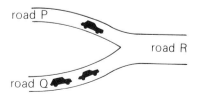

road P

road R

road Q

Imagine also three pipes through which water is flowing at a rate of 1 litre per second, 2 litres per second and 3 litres per second, as shown below.

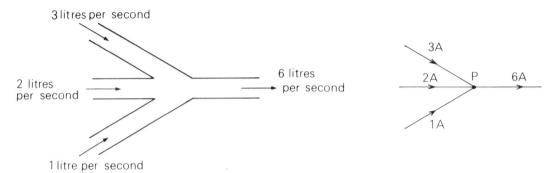

3 litres per second

2 litres per second

1 litre per second

6 litres per second

3A

2A P 6A

1A

The rate of flow of water through the pipe after the smaller ones have joined is 6 litres per second. In the second figure, currents of 1 ampere, 2 amperes and 3 amperes reach the point P and the current leaving is therefore 6 amperes.

We have mentioned cars moving along roads and water flowing through pipes. It is therefore an obvious question to ask what is flowing when there is an electric current in a wire. You cannot see an electric current. Our experiments have in fact merely shown us what an electric current does. Some of you may have heard that there are electrons travelling along the wire but for the present we will leave this question and concentrate on what an electric current does.

Questions for homework or class discussion

1. All the circuits use similar cells and similar lamps. In Figure i the current is 0.2 amperes. For each of the other circuits, say whether the current will be zero, between 0 and 0.2 A, 0.2 A, or greater than 0.2 A. Give a reason for each answer.

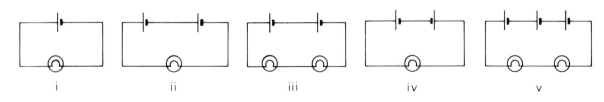

i ii iii iv v

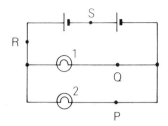

2. Two cells are connected across two similar lamps as shown. The current passing through an ammeter at P is 0.2 amperes.

a. What is the current at Q?

b. What is the current at R?

c. What is the current at S?

d. Redraw the circuit and insert a switch (which you should mark with the letter A) which will switch both lamps at once, and a switch (marked with the letter B) which will switch off only the lower lamp.

3. What is represented by the symbols labelled a, b, c, d and e? Which pair of lamps P and Q, or R and S, are in series? Which pair are in parallel?

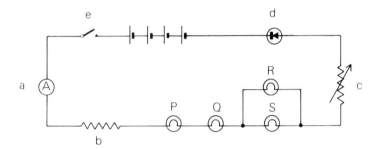

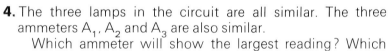

If the lamps P, Q, R and S are all similar, would P be more bright, equally bright or less bright than Q? Would P be more bright, equally bright or less bright than R?

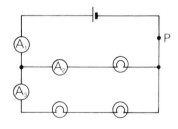

4. The three lamps in the circuit are all similar. The three ammeters A_1, A_2 and A_3 are also similar.

Which ammeter will show the largest reading? Which ammeter will show the smallest reading?

If another ammeter were inserted in the circuit at the point marked P would it read more, less or the same as ammeter A_1?

5. In each of the circuits below, the cells are all similar and the lamps are all similar. When one of the cells is put across one of the lamps, the lamp glows with normal brightness and the current that flows through the lamp is 0.2 A.

a. What is the reading of ammeters A_1 and A_2?
b. In the second circuit, the cell is across two lamps in parallel.

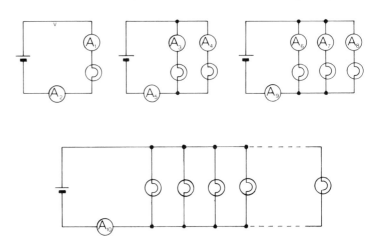

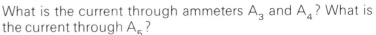

What is the current through ammeters A_3 and A_4? What is the current through A_5?

c. In the third circuit, the cell is across three lamps in parallel. What is the current through A_6, A_7, A_8 and A_9?

d. The final circuit is a way of showing that the cell is across a large number of lamps though it is not clear from the diagram exactly how many. If there were ten lamps, what would you expect the current to be through A_{10}? Do you think this is what the reading would be if you did this experiment? To find out, the best thing is to do Experiment 16.4 to see for yourself.

Experiment 16.4　Lamps in parallel
Set up the circuit shown on the left. What is the reading of the ammeter?

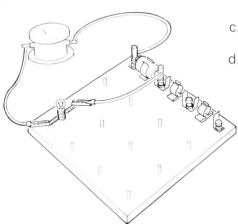

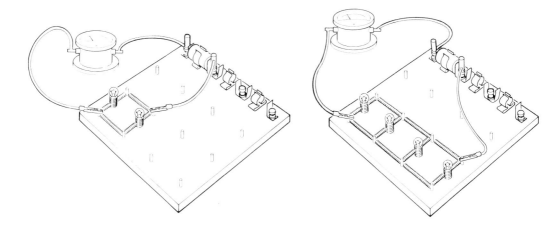

Put two lamps in parallel as shown in the first diagram. What is now the reading of the ammeter? Is it exactly double the reading when there was only one lamp?

Then put three lamps in parallel. Finally try four lamps in parallel. Each lamp has the cell connected across it, but they do not glow with the full normal brightness, and the current is not quite four times what it was originally.

If you like, you can try more lamps in parallel using two circuit boards. You will find you cannot go on getting more and more current from your cell: there seems to be a maximum current it will give.

If an ammeter is connected straight across a cell without any lamps in the circuit, it will be found that the current is about 1.2 A. This is therefore the maximum current the cell will give. There is of course some resistance in the connecting leads and in the ammeter itself, but the main resistance which restricts the current to a maximum value is the resistance inside the cell itself. This is usually called the *internal resistance* of the cell.

How long will a cell last?

Consider the three circuits below. The lamp will glow brightest in circuit i, but there are two cells. The lamp will glow with normal brightness in circuit ii. In circuit iii there are two cells, but three lamps each of which will glow with less than normal brightness. In which case will the cell or cells run down most quickly?

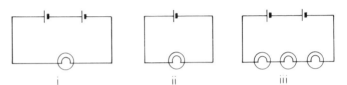

In circuit ii the current may be about 0.2 A, but in circuit i the current is greater, probably over 0.3 A. It is important to remember that this current flows all round the circuit, which includes the two cells. Thus over 0.3 A flows through each cell in circuit i and only 0.2 A through the cell in circuit ii. For this reason both the cells in circuit i will run down faster than the cell in circuit ii. The current is least in circuit iii, so those cells will last longer.

In other words, to decide which cell will last longest, you merely have to find which cell has the smallest current flowing through it.

Cells in parallel

Until now we have considered cells only in series, as in the following circuits.

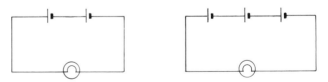

In the first circuit, the lamp glowed brighter than normal and in the second circuit very much brighter.

What would be the effect of putting similar cells in parallel, as shown below?

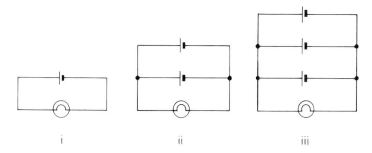

i   ii   iii

As far as the lamp is concerned, it would be just the same as putting one cell across it. In each case, it would merely glow with normal brightness. The difference would be the amount of current taken from each cell. In circuit i, the current through the lamp might be 0.2 A, and 0.2 A would flow through the cell. In circuit ii, the current through the lamp would still be 0.2 A, but only 0.1 A would flow through each cell. What would be the current through each cell in circuit iii?

The advantage of putting cells in parallel would be that they would last longer.

Experiment 16.5 Further experiments with circuit boards

Set up this circuit. Fix two crocodile clips at the points X and Y so that different things can be put into the circuit between them. Try each of the items in the list below. First guess what will happen to the reading on the ammeter and then see if your guess is correct.

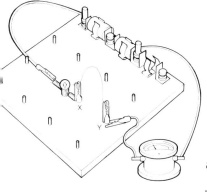

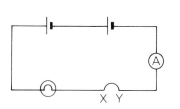

a. A piece of copper wire and then the same length of very much thinner copper wire.

b. A strip of copper and then a strip of paper.

c. A short length of eureka wire (resistance wire) and then a longer piece of the same wire.

d. A length of eureka wire and then eureka wire of the same length, but larger in diameter.

e. A fixed resistor and then a rectifier (in each case see what happens when they are turned round the other way).

Experiment 16.6 The heating effect of a current

Set up the circuit using three cells, an ammeter and a variable resistor. Between the two crocodile clips stretch a few strands of steel wool.

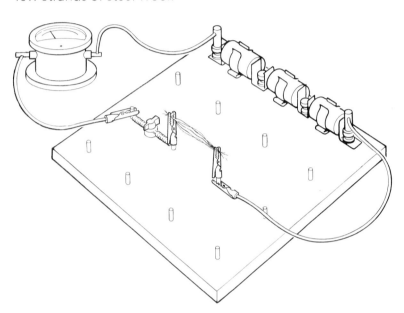

Start with a small current. Gradually increase the current by reducing the resistance. The steel wool will be found to get hot. The greater the current the hotter it will get.

Try blowing hard on the wire so that it is cooled. You should notice that the ammeter reading increases. In other words the resistance depends considerably on temperature: the greater the temperature the greater the resistance.

If only one strand of steel wool is used, it may get so hot that it melts and no current flows. This is the way in which a fuse works. To prevent too great a current causing damage to the wiring, a fuse is included in a circuit. When the current is too great, the wire melts, the circuit is broken and current can no longer flow. The symbol for a fuse in a circuit diagram is:

Fuse wire

Experiment 16.7 Testing fuses

The photograph on the left shows typical fuse wire, and the drawings the small cartridge fuses used in plugs and apparatus. The cartridge fuse has a fine wire inside which melts as soon as the current exceeds the stated value.

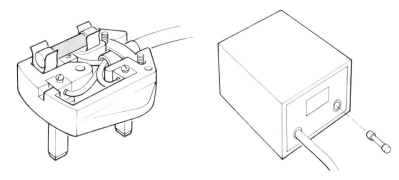

To test one of these cartridge fuses set up the circuit below, fixing a $\frac{1}{4}$ ampere fuse between two crocodile clips as shown.

Start with a small current. Then decrease the resistance, watching the ammeter reading until the fuse blows.

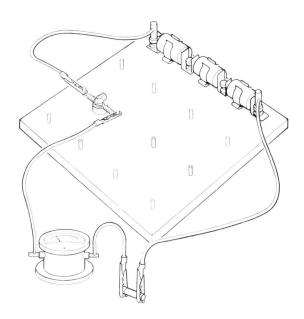

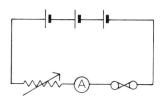

Homework assignments

1. Make a list of ten things worked by electricity in your home or in your school.
2. Collect photographs, advertisements and cuttings from papers which show electricity in action.
3. Read the topic book *Electric currents*, pages 7–11, and prepare a short talk on the heating effects of an electric current.
4. Read *Electric currents*, pages 12–19, and prepare a short talk on electric lighting.
5. Fuses are very important to protect apparatus and for safety purposes in your home. Read *Electric currents*, pages 31–35, and prepare a short talk on fuses.
6. Read *Electric currents*, pages 35–38, and then write a paragraph in answer to each of these three questions.
   a. What is meant by earthing?
   b. Why are there usually three wires connected to an electric fire?
   c. What is a lightning conductor?
7. Read *Electric currents*, pages 39–46, and either write an essay or prepare a talk on the electric wiring in a house.
8. Read *Electric currents*, pages 47–55, and prepare a talk on switches and automatic control.

Electrical pressure

In order to drive water round a water circuit, it is necessary to have a pump in the circuit. The pump causes a difference in water pressure between the two sides of it, and it is this difference in pressure which causes a current of water to flow.

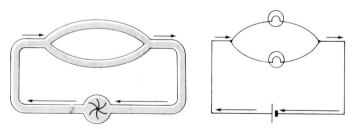

It is much the same in an electric circuit. A cell causes a difference in electrical pressure, and it is this difference in pressure which causes an electric current to flow. If the electrical pressure difference is small, only a small current will flow. But if the pressure difference is increased, the current will be bigger. The pressure difference is the *cause* and the current that flows is the *effect* which results from it.

Electrical pressure difference is usually measured in *volts*. One cell is about 1.5 volts. If two cells are put in series, the pressure difference is greater and will be 3.0 volts. Three cells in series give 4.5 volts, and so on.

In the water circuit illustrated we could have measured the difference in pressure across the pump by connecting tubes on either side of it. The difference in pressure would be shown by the difference in height of the water in the two tubes.

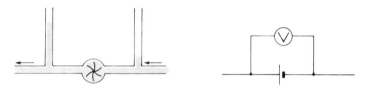

In a similar way electrical pressure can be measured across a cell by connecting a *voltmeter* across it.

Experiment 16.8   A voltmeter as a cell counter

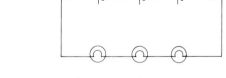

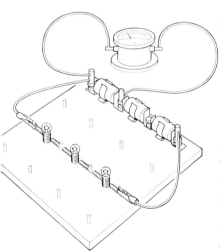

Connect three cells across three lamps on your circuit board. Then connect your voltmeter across P and Q and notice the reading. Then connect between Q and R, then between R and S. The readings should all be the same, about 1.5 volts.

Then connect the voltmeter between P and R, and between Q and S. This is the reading for two cells. Finally connect between P and S.

Questions for homework or class discussion

1. A lamp connected to a cell as in Figure i glows with normal brightness.

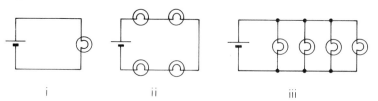

What would happen to the brightness if four similar lamps were connected to the cell as in Figure ii?

What would happen to the brightness if the four lamps were connected as in Figure iii?

In which of the three cases, Figures i, ii or iii, would the cell run down most quickly? Give the reason for your answer.

2. In these circuits, all the cells are similar and all the lamps are similar. Assume in each case that all connections are made with very good conducting wire.

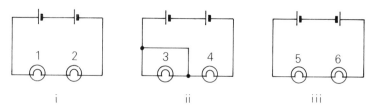

In circuit (i) the lamps 1 and 2 glow with normal brightness. State whether the lamps 3, 4, 5 and 6 will be brighter than normal, normal brightness, less than normal brightness or will not glow at all.

In which case will the cells run down most quickly? Give a reason for your answer.

3. Two similar model boats A and B are fitted with similar electric motors to drive the propellers. Similar cells are used to drive the motors, but the circuits used are as shown below.

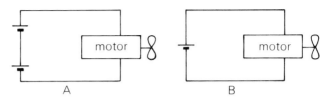

Which of the two boats is likely to go faster? Give a reason for your answer.

In which of the two boats will the cell or cells run down faster? Give a reason.

4. The battery X is used to light the two lamps P and Q. It is found that P glows more brightly than Q.

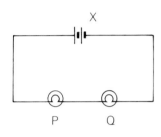

Someone suggests that this is 'because the current decreases as it moves round the circuit'. How would you show this to be either correct or incorrect

a. if you had a suitable ammeter?

b. if you had no other apparatus?

If you think the suggestion was incorrect, give a more likely explanation.

5. Describe what will be the effect of decreasing the resistance of the variable resistor in each of the following:

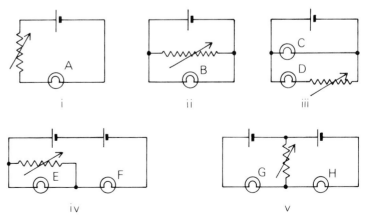

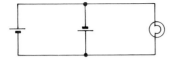

6. Explain why the circuit on the left is a 'fool's circuit'.

7. Describe what will be the effect of closing the switch in each of the following circuits:

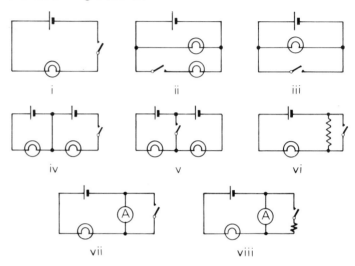

8. The following circuits include rectifiers. Which lamps (A, B, C, etc.) will light?

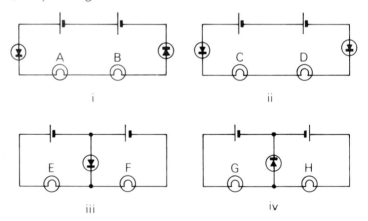

212

9. A closed box has four terminals, A, B, C and D, and two similar lamps, X and Y, screwed into sockets in it.

When a cell is connected to terminals A and C, lamp X lights brightly. When the cell is connected to A and D, nothing happens until B and C are connected by a piece of copper wire, when both lamps light dimly.

Draw a diagram to show the connections inside the box.

How would you connect the cell to make lamp Y light brightly without lamp X lighting?

How would you get both lamps to light brightly?

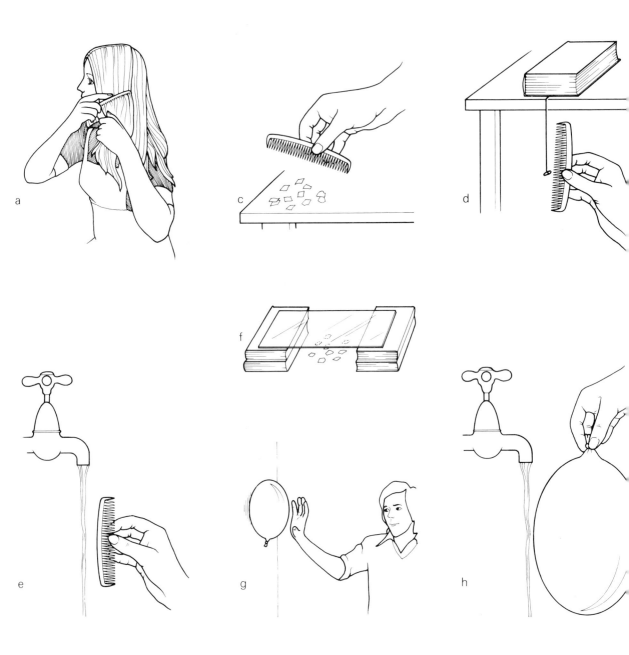

a

c

d

e

f

g

h

Chapter17 # Electric charge and electric current

You have probably done experiments at home using a comb to lift pieces of paper. You might also try the following. They will work best under very dry conditions.

a. Try combing your hair with a nylon comb in front of a mirror in a darkened room. What can you hear? Can you see anything?

b. Pull off a nylon shirt or nylon sweater in the dark. Do you hear strange crackles or see flashes of light?

c. Comb your hair or rub the comb on a woollen sleeve. Hold the comb over some small pieces of paper on a table.

d. Attach a grain of puffed rice to a fine nylon thread. Support one end of the thread under a book and let the puffed rice hang free. Comb your hair and bring the comb near the puffed rice.

e. Turn on a water tap to get a fine jet of water and hold the comb near the jet.

f. Place a piece of plastic or glass on top of books as shown. Put some shapes made from tissue paper under the plastic. Rub the plastic or glass with a duster. What happens to the tissue paper shapes?

g. Blow up a balloon and rub it on a carpet. Then put the balloon against the wall of a room or touch the ceiling with it. Does it stay there when you let go?

h. Rub a balloon on a carpet and hold it near to a jet of water (as in e).

These are all strange effects, and scientists find the best way to explain them by introducing the idea of *electric charge*. We cannot see electric charge, any more than we can see an electric current, but we believe it to exist because of its effects. We say that when a comb, a balloon or a plastic sheet is rubbed that it has become *charged*.

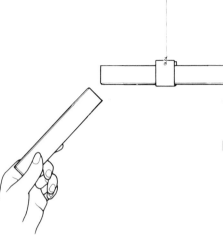

Experiment 17.1 Evidence for two kinds of charge

For this investigation you will need strips of polythene and cellulose acetate, together with a duster or large handkerchief. Try the following experiments.

a. Rub both a polythene strip and a cellulose acetate strip to show they can both pick up small pieces of paper. This shows that both strips become charged when rubbed.

b. Use a wire stirrup (or one made from a strip of paper) to hang up a polythene strip so that it hangs freely from a length of nylon thread. Rub both ends of the strip with the duster.

   Also rub another polythene strip. Bring *one* of the ends close to *one* end of the suspended strip. What happens? Try the other end. What happens?

   As both rods are of the same material and were rubbed with the same duster, they must be charged the same. You notice that, whichever end is used, the charged rods repel each other.

c. Now repeat all of Experiment b, but use instead two cellulose acetate strips. Once again whichever ends you use, the charged rods repel each other.

d. Finally hang up a polythene strip. Rub it with the duster. Then rub a cellulose acetate strip with the duster and bring it near one end of the polythene strip. This time they attract each other.

   Thus it appears that there are two kinds of charge, one is found on a rubbed polythene rod, the other on a cellulose acetate strip. It also appears that like charges repel each other, unlike charges attract. For convenience it is usual to refer to these two kinds of charge as *positive* and *negative*. (In fact the polythene strip is negatively charged, the cellulose acetate positively charged when rubbed with the duster.)

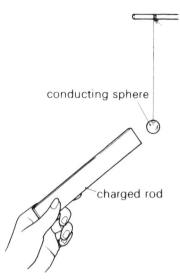

conducting sphere

charged rod

Experiment 17.2 Investigation of charges using light spheres

As well as the polythene and cellulose acetate strips, you will need two light, conducting spheres suspended from fine nylon thread: these can be very light expanded polystyrene spheres coated with aluminium paint to make them conducting. They can be fixed to the nylon thread with a small piece of Sellotape.

a. Hang up one of the conducting spheres. Bring a charged polythene strip near to it. What happens?

b. Let the sphere touch the strip so that it is charged by it. Remove the strip and then bring it slowly towards the sphere. What happens this time? Can you explain this?

c. Bring up a charged cellulose acetate strip near to the charged sphere. What happens this time?

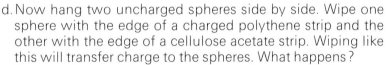

d. Now hang two uncharged spheres side by side. Wipe one sphere with the edge of a charged polythene strip and the other with the edge of a cellulose acetate strip. Wiping like this will transfer charge to the spheres. What happens?

e. Repeat d, this time wiping both spheres with the same charged rod. What happens?

The van de Graaff generator

We have been producing charges by rubbing. An ingenious way to produce large charges was invented in 1931 by van de Graaff and a photograph of a large van de Graaff generator is shown on the left. We use a small one in school laboratories: it can build up very large charges on the big collecting sphere at the top when the machine is driven either by hand or by an electric motor.

The Aldermaston 6 million volt van de Graaff generator

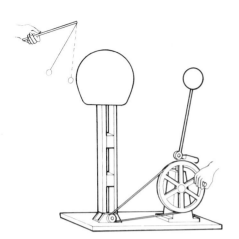 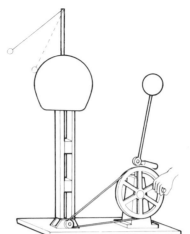 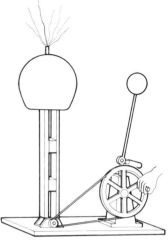

Experiment 17.3　Experiments with a van de Graaff generator

a. Hang a small conducting sphere near to the big collecting sphere of the generator. Let the small sphere touch it and then, as like charges repel, the small sphere will be pushed away.

b. Fix a rod with a small conducting sphere on the top of the generator. Again the small sphere will be repelled.

c. Fix a 'head of hair' on the top of the generator. Charge it up and watch the repulsion.

d. Hold your head near the generator and feel the effect on your hair.

e. If someone in rubber boots stands with a hand on the conducting sphere while it is being charged up, he will also be charged up. Charge can then be dramatically transferred to someone else who touches his other hand!

f. Finally, bring an earthed sphere near to the collecting sphere so that a spark can pass from it to earth.

Conduction of electricity in liquids

In previous chapters, we found that certain solids were conductors of electricity, whereas other substances were insulators. We shall now investigate whether liquids are conductors or insulators.

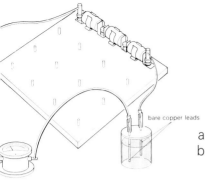

bare copper leads

Experiment 17.4   Investigation of liquids as conductors

Connect three cells in series with an ammeter and two bare copper leads as illustrated. The bare ends of the copper wires should hang down into the beaker, but they must not touch each other. The beaker will be filled in turn with the various liquids to be investigated.

a. First try distilled water. Does any current flow?
b. Add a little common salt to the water. Stir the solution and see what happens. If you see anything happening at the copper wires, break the circuit temporarily to see whether it is due to the current or not.
c. Empty the beaker and wash it out. Refill it two-thirds full with distilled water and this time add a little dilute sulphuric acid and see what happens. Switch off the current. Does the effect stop? This shows that it is due to the current and not to some chemical reaction in the solution.
d. Repeat with fresh distilled water to which sugar has been added. Alternatively, try paraffin on its own.
e. Finally try tap water.

Electrolysis

You found in the last experiment that very pure water (distilled water) did not conduct electricity, but a very small amount of impurity made it conduct. You have doubtless heard how dangerous it is to have any mains-operated electrical appliance in a bathroom: your dirty bath water is a long way from being pure and is a good conductor of electricity.

You will have noticed that when a current passed through the slightly acidified water it produced certain chemical changes. In all scientific work it makes it easier to discuss details if we give names to parts of the apparatus. A solution through which the current passes is called an *electrolyte*. The two electrical contacts immersed in the electrolyte are called *electrodes*: the one connected to the positive terminal of the battery is called the *anode* and the one connected to the negative terminal the *cathode*. The whole process is called *electrolysis*.

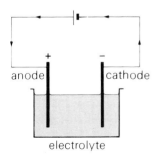

anode   cathode

electrolyte

The chemical changes which occur during electrolysis are often complicated. They depend both on the electrolyte and on the nature of the electrodes. We shall not consider electrolysis in detail here, except for the interesting cases shown in the following experiments.

Experiment 17.5　Electrolysis of copper sulphate solution

For this experiment use the same circuit as in Experiment 17.4, but for electrodes use two strips of copper foil slipped down the inside of the beaker and bent over at the top. They can be connected into the circuit with crocodile clips. Make certain they do not touch each other.

Fill the beaker with some strong copper sulphate solution. Switch on the current. Let it flow for several minutes and then look at the electrodes to see if there is any difference. You should see that clean, pure copper has been deposited on the cathode, and you could confirm this by measuring its mass both before and after the current flowed. Similarly you could find that the anode has lost mass.

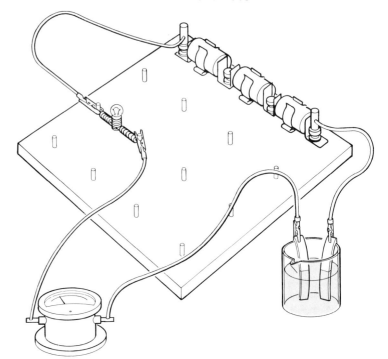

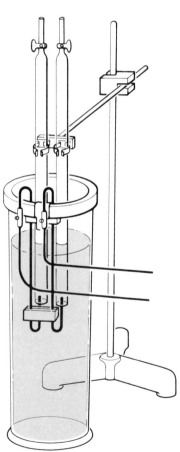

Experiment 17.6   Copper plating

Use the same circuit as in the previous experiment, but instead of a cathode made of copper foil try other articles, like a brass screw, a paper clip or a coin. (It is better to avoid objects made of zinc or iron for these will react of their own accord with the solution.) Fill the beaker with strong copper sulphate solution and pass the current.

Experiment 17.7   Electrolysis of slightly acidified water

In experiment 17.4 you saw that a few drops of dilute sulphuric acid added to distilled water made it conduct. You will have noticed some bubbling at the electrodes, and the object of this experiment is to investigate that further.

The large container has distilled water in it with a little sulphuric acid added to make it conducting. A current of about 0.5 ampere is passed through it and bubbles are again seen at the electrodes. The gases given off are collected in the glass tubes held over them. It can be shown that the gas given off at the anode is oxygen and that at the cathode is hydrogen. A glowing wood splint bursts into flame when put into oxygen, whereas hydrogen in a test tube explodes with a pop when it is lit.

Experiment 17.8   Lead tree demonstration

This demonstration is a beautiful illustration of electrolysis. This time the electrolyte is lead acetate and the electrodes have been bent as shown. When a small current (about

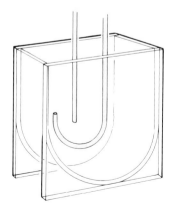

50 mA) is passed, a tree of lead crystals will grow. It looks particularly good if the cell is put in a slide projector so that the electrodes and the tree are in focus on the screen.

What happens when the current is reversed?

How does a current flow in a liquid?

How can we explain electrolysis? It is outside the scope of this book to discuss it in detail and you will learn about it in your chemistry course, but one thing seems clear from these experiments. Something is moving through the electrolyte. In one experiment copper was deposited at one of the electrodes, in another lead was deposited, whereas in the acidified water experiment oxygen was deposited at one electrode and hydrogen at the the other.

A possible explanation might be that something charged moves through the liquid. Of course this is only a guess, but you will find that other experiments, particularly in chemistry, support this idea. We know already that opposite charges attract each other. These *ions,* as they are usually called, have charge, and the negative ions move to the positive anode and the positive ions to the negative cathode. It is these moving ions that are responsible for the current through a liquid.

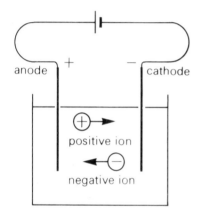

anode
+
cathode
−

positive ion

negative ion

Homework assignment

Electrolysis has important practical applications in the world. It is used for *electroplating* and for *refining of metals.* Read the topic book *Electric currents,* pages 26–30, and write a short essay or prepare a short talk on one of these two topics.

Conduction of electricity in gases

We have seen already that some solids are conductors and some insulators. We found the same with liquids: some conduct, others do not. What about gases?

In your experiments with circuit boards, you found that any gap in the circuit stopped the current from flowing: air does not conduct electricity. Think how awkward it would be if it did: mains sockets would be short-circuited and so would fuses. Batteries would not last long.

However, you have now seen sparks produced by a van de Graaff generator. In the same way when a cloud becomes highly charged, a lightning flash may occur: a huge spark passes through the air either from one cloud to another, or from the cloud to the earth.

What makes it possible for a current to flow through the air? Perhaps *ions* are present, in the same way that we suggested positive and negative ions might carry the current through a liquid. The next experiment gives some evidence for ions in the air.

Lightning sparks

Experiment 17.9　Ions in the air produced by a candle

Two metal plates, about 5 cm apart, are set up above a lighted candle. One of the plates is connected to the large sphere of the van de Graaff generator, the other to earth. This means that when the generator is operating there will be several thousand volts across the plates.

A strong light source casts a shadow of the flame on a screen.

What happens to the shadow when the voltage is applied? It should be seen that the flame divides into two parts, one towards the positive plate and one to the negative plate. It looks as though the candle produces ions in the air, negative ions which move one way and positive ions the other.

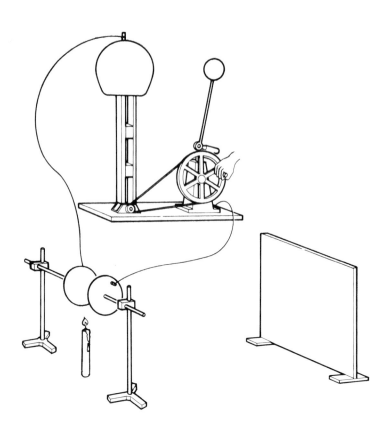

Moving charge and current

We have learnt in this chapter about electric charge and we already know about electric currents, but we have not yet shown a relationship between them. For that experiment we need to use a new instrument called a *galvanometer*. This is a detector of very small electric currents. (It is named after an Italian professor of anatomy, Luigi Galvani, who in 1780 used the legs of dead frogs to detect electric currents: the legs twitched when a current passed.)

You can show how sensitive a galvanometer is by putting it in series with a cell and a person. With an ordinary ammeter you would not be able to detect any current. (Try it and see.) The resistance of a person is so high that with only one cell the current is very small indeed, but it can be detected with a galvanometer.

This experiment shows that the galvanometer reading changes when a very small current passes through it.

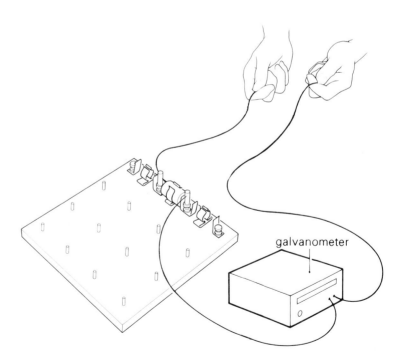

galvanometer

Experiment 17.10 A current and moving charge

A table-tennis ball is coated with graphite (Aquadag) to make it conducting and it is hung by a very long thread between two metal plates as shown. One plate is connected to the collecting sphere of a van de Graaff generator. The other plate is connected to one terminal of the galvanometer, the other terminal of which is connected to the earth connection on the van de Graaff generator.

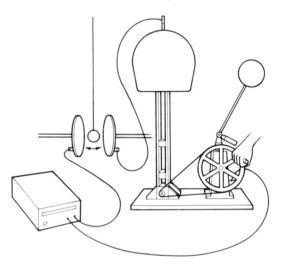

The van de Graaff generator is set going so that the large sphere is charged. The table-tennis ball oscillates between the two plates carrying charge across from one to the other, and the galvanometer records a current.

This experiment shows us that moving charge causes a current. From now onwards we can assume that an electric current is the movement of electric charge. Of course the experiment has not shown whether it is positive charge moving one way or negative charge moving the other. A decision on that will have to wait until later in your physics course.

Questions for class discussion

1. A thin piece of copper wire can safely be used to connect a lamp to a cell. Yet the same wire when connected across a mains socket would immediately produce a disaster. Explain what happens in both cases and account for the difference.

2. If a wire is connected straight across the terminals of a 12-volt car battery, a current of 50 amperes might flow. A current of 1 ampere is sufficient to kill a man. If you touch the two terminals of a car battery you will not feel even the slightest shock. Explain this.

3. You can get rid of the charge on a polythene or cellulose acetate strip by pulling the strip through the flame of a Bunsen burner or a candle. Can you explain why it loses its charge?

4. Six conducting balls are hung from insulating threads. Ball 1 is positively charged.

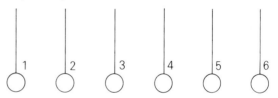

In succession, different balls are brought near to each other and various effects are observed.

a. When ball 2 is brought near ball 1, it repels it.
b. When ball 3 is brought near ball 2, it repels it.
c. When ball 3 is brought near ball 4, it attracts it.
d. When ball 4 is brought near ball 5, it repels it.
e. Ball 6 is found to be attracted to both ball 3 and ball 4 when brought near to them.

Decide whether the balls 2, 3, 4, 5 and 6 are positively charged, uncharged or negatively charged. Give your reasons.

5. You have seen a van de Graaff generator in operation.
a. Why does it not work well when the weather is damp?
b. Why is an electric shock received from a van de Graaff generator (about 50 000 volts) less dangerous than one received from the domestic mains supply (230 volts)?
c. Why do you not receive a shock if you stand on an insulating platform and charge yourself up by holding the live sphere?
d. What do other people in the room notice about your appearance as you are being charged up?
e. The machine is now switched off and you shake hands with your friend who has been standing on the floor beside you. Why do you both receive a shock?

Revision test D1

1. A spring has an unstretched length of 10 cm as in Figure i. When a mass of 200 g is added, its weight stretches the spring 2 cm to a new length of 12 cm.

a. What will be the length if only 100 g is attached to the spring instead of 200 g?

b. A second and similar spring is attached to the end of the first, as in Figure ii. What will be the *total length* if 200 g is attached to the lower end?

c. The two springs were put side by side and joined with a light rod, as in Figure iii. What will be the *length* of each spring if 200 g is attached at the centre point P?

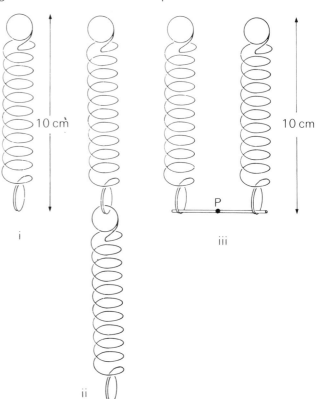

2. Forces are measured in newtons. The downward force due to gravity on a mass of 1 kg at the surface of the Earth is about 10 newtons.

a. What would be the downward force on a mass of 2 kg?

b. And on a mass of 3 kg?

c. On the Moon gravity is only one-sixth of the value on the Earth. What would be the downward force, measured in newtons, due to gravity on a mass of 3 kg on the surface of the Moon?

d. 1 kg of butter has the same *mass* (equal to 1 kg) whether it is measured on the Earth or on the Moon. Is the *weight* of the butter on the Moon equal to, less than or more than the weight of the butter on the Earth?

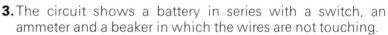

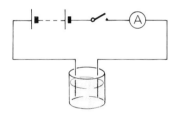

3. The circuit shows a battery in series with a switch, an ammeter and a beaker in which the wires are not touching.

The simple switch is made from two metal strips fixed to a base. What material would be suitable for the base? What material would be suitable for the connecting wires?

Tap water, distilled water and copper sulphate solution are separately and in turn poured into the beaker to complete the circuit. Which gives the highest reading on the ammeter when the switch is closed?

The two wires are removed from the beaker and are first joined with a piece of thick iron wire to complete the circuit. Then a very thin piece of iron wire of the same length is used instead of the thick iron wire. With which piece of iron wire will the current be larger when the switch is closed?

4. Certain makes of jam are sold in pots with a top which you cannot pull off by hand. There is a printed instruction saying 'Pierce with a pin to release vacuum'. What do you think happens when the top is pierced? Explain why it is difficult to remove the top before it is pierced.

5. Explain the reason for the following.

a. The handle of a door is usually placed a long way from the hinge.

b. When a lamp, a variable resistor and a cell are connected in series, the brightness of the lamp can be changed by moving the knob on the variable resistor.

c. When a bubble of air is introduced at the bottom of a barometer, it will increase in size as it rises through the mercury.

Revision test D2

1. You are given two similar cells and two similar lamps, such that if one cell is connected across one lamp it glows with normal brightness.

Draw circuit diagrams to show how you would connect one or both cells to the lamps so that (a) both lamps glow with normal brightness, (b) both lamps are dimmer than normal, (c) both lamps are brighter than normal.

2a. What is a fuse? Explain its purpose.

b. What would be the difference between a 1 A fuse and a 5 A fuse? Explain the reason for the difference.

c. Explain how you would test a fuse rated at $\frac{1}{4}$ ampere using a circuit board. You have three cells, an ammeter and a variable resistor. A circuit diagram will help your explanation.

3. A boy has a microbalance set up correctly with the pin near the top of the straw at A, but he finds that the pointer moves only 1 mm when he uses it to weigh a hair and he would like it to move about 5 mm. His friends make various suggestions. Say what the effect of each would be.

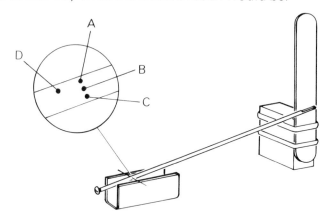

a. 'Cut off a little from the pointed end of the straw.'
b. 'Move the pin along to D.'
c. 'Move the pin to B, just *above* the centre of the straw.'
d. 'Move the pin to C, just *below* the centre of the straw.'
e. 'Use a thicker pin in hole A.'

Which of these suggestions do you think is best?

The boy finds that it takes too long for the balance to settle down so he tries fastening a piece of paper to the pointed end of the straw. He tries putting it both vertically and horizontally as shown.

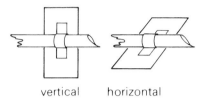

vertical   horizontal

f. Which direction do you think would be better? Give a reason.

g. Would he have to make any changes at the other end of the balance? If so, what changes?

4. The diagram illustrates a mercury barometer correctly set up.

a. Which of the distances A, B, C or D measures the atmospheric pressure?

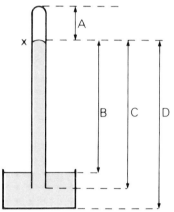

b. More mercury is added to the container so that the level in it rises by 1 cm. What will happen to the mercury level in the barometer tube at X?

c. If a bubble of air is allowed to rise up the tube towards X, what will eventually happen to the level of the mercury at X?

5. If smoke is inserted into a cell, Brownian motion can be seen when the smoke particles are viewed through a microscope. What does this motion look like?

How can you explain the motion?

Revision test D3

1. A lamp connected across a single cell, as in Figure i, glows with normal brightness.

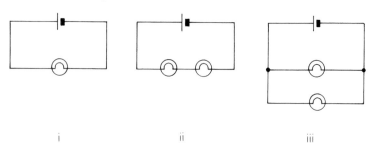

i ii iii

a. What happens to the brightness if two similar lamps are connected to the cell as in Figure ii?
b. What happens to the brightness if the lamps are connected as in Figure iii?
c. In which of the three circuits shown would the cell last the longest? Give a reason for your answer.

2. The diagram shows a simple circuit in which a single cell lights two similar lamps with normal brightness.

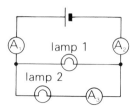

If the ammeter A_1 reads 0.5 amperes, what will A_2 read? What will ammeter A_3 read?

Draw a circuit diagram similar to this, but omit the ammeters and include a variable resistor to dim both lamps at once and, a switch to turn out lamp 1 without affecting lamp 2.

3. The diagram shows four tubes of glass connected at the top to the same pipe which is connected to a pump. The open ends of the tubes are well below the surface of the water in a large open bowl. Another tube E dips below the surface of the mercury in another bowl.

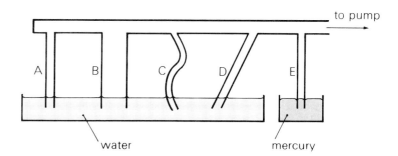

The pump is used to produce a partial vacuum in the top tube and the water rises half way up tube A.

a. What happens to the water level in tube B?
b. What happens to the water levels in tubes C and D?
c. What happens to the level of mercury in tube E?
d. How high would the water tubes have to be if the pump produced a nearly complete vacuum and the water were not to flood the top pipe?

4. A pop-gun can be made from a piece of thin-walled brass tubing T and a solid pusher rod R with the same relative size shown in the diagram. The ends of the tube are stopped up with two small, closely-fitting pieces of potato, A and B.

a. If R is pushed quickly into the tube, A will shoot out of the tube before B reaches it. Why does this happen?
b. As it does so it emits a loud 'pop'. Why?
c. B remains in the tube, more or less where A was at first. Why does it not come out?
d. If R is pushed very slowly, A will usually remain where it is. Why?

5. Explain each of the following as fully as possible.
a. It is uncomfortable to walk over gravel with bare feet.
b. A wheelbarrow enables you to move a load which is too heavy for you to lift unaided.
c. A block of steel will sink in water, but a steel needle can be made to float.
d. A very small oil drop placed on water in a large tray spreads out into a thin film on the water's surface, but stops spreading before it reaches the edge of the tray.

Revision test D4

1. An electric current is able to do a number of things and three of these are mentioned below. Describe an experiment you would do to demonstrate each of them. In each case, mention what apparatus you would use and draw the circuit diagram.
a. An electric current can be used to produce heat.
b. An electric current has an effect on a small compass needle.
c. An electric current can be used to bring about chemical changes.

2. A battery, a lamp and a switch are connected in series. When the switch is closed, the lamp fails to light. List *three* possible reasons. How would you test each possibility?

3a. A rectangular block of metal has sides 3 cm, 4 cm and 5 cm respectively. What is its volume?
b. If the mass of the block is 480 g, what is the density of the metal?
(Be careful to give the units in both the above.)
c. Write down the following in order of increasing density, putting the least dense first: lead, air, water, foamed polystyrene.

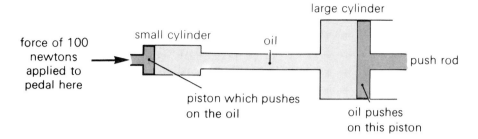

force of 100 newtons applied to pedal here

small cylinder

oil

large cylinder

push rod

piston which pushes on the oil

oil pushes on this piston

4. The diagram shows a simple mechanism to operate the brake on a motor car. The area of the small cylinder is 2 cm² and of the large cylinder 10 cm². The system is filled with oil.

a. If the pedal pushes against the piston in the small cylinder with a force of 100 newtons what is the pressure exerted on the oil?

b. What is the force exerted by the oil on the piston in the large cylinder?

c. If the push rod has to move 1 cm to operate the brake, how far must the piston in the small cylinder move?

d. Explain what would happen if there were a bubble of air in the system.

e. Suggest an advantage that such a system might have over a lever operated system.

5 a. A tennis ball and a ball of Plasticine are both dropped on to hard ground. What kind of energy do the balls have when held before being dropped?

b. What kind of energy do the balls have just before they hit the ground?

c. Why does the tennis ball bounce while the Plasticine ball does not?

d. Why does the tennis ball not rise as high again as the point from which it was released?

e. When the tennis ball has eventually come to rest on the ground, has it more or less energy than before it was dropped?

f. If you think it has more, explain where the extra energy has come from; if you think it has less, explain what has happened to the energy it has lost.

Descriptive revision test D5

1. You can find the volume of a rectangular block by measuring the length, breadth and height, and multiplying them together. But you cannot do this with an irregular granite rock. Describe an experiment you might do to find the volume of such a granite rock.

2. Draw a sketch of a drinking-straw microbalance.

How would you use the microbalance to find the mass of a dead insect? You can assume you have a pile of graph paper and a laboratory balance capable of measuring masses up to 1 kg.

3. The magnetic effect of an electric current can be used to measure an electric current. Describe, with a diagram, how a current balance is made.

4. Explain how a fuse acts as a safety device in electrical apparatus.

5a. Describe how you would set up an experiment to observe Brownian motion, the irregular motion of small smoke particles in air.

b. Describe what you would expect to see.

c. What conclusions can you draw from such an experiment?

6. You want to find the diameter of a piece of copper wire, but it is too thin for you to measure directly. Someone suggests you should wind it round a pencil twenty times. Explain how this might help you to measure the diameter.

Chapter 19 **A further look at energy**

Measuring energy transferred

In Chapter 13 we were introduced to energy in many different forms. There were many examples of energy being transformed from one form to another. In this chapter we will discuss how the energy transferred is usually measured.

You are certainly familiar with the word *work*. We speak about doing work digging a garden or lifting sacks of potatoes, we may say we are doing some hard work reading a history book. But this everyday use of the word *work* is a bit vague and scientists prefer to give a precise meaning to it. They say that work is a measure of the amount of energy transferred.

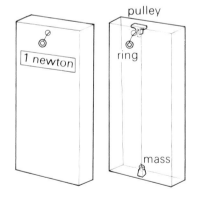

You are already familiar with the forces demonstration box. If you pull on a ring, you can feel a force of 1 newton. You will also know how the box is made: when you pull with a force of 1 newton, a mass inside the box is raised. If you pull the force through a distance of 1 metre, the mass is raised 1 metre. In other words chemical energy in your body has been transformed into uphill energy of the mass. Another name for uphill energy is *gravitational potential energy*, or more shortly *potential energy*, and we will use this in this chapter. The more usual scientific name for motion energy is *kinetic energy* and we will also use that in future.

Work was done in moving the force of 1 newton through 1 metre and we call this amount of work 1 *joule*. This measures the amount of energy transferred from chemical energy in your body to potential energy of the mass. (The abbreviated way of writing 1 joule is 1 J.)

work = force × distance

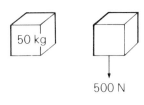

500 N

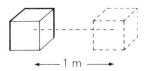

← 1 m →

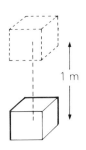

1 m

We have defined work as the energy transferred and have said that it is calculated from the force applied multiplied by the distance moved. But we must be careful about what we mean by the distance moved.

Imagine a large block of ice with a mass of 50 kg resting on a flat glass surface. It will be very easy to push the block sideways through a distance of 1 metre since there will be only a small frictional force of a few newtons. But the downward force due to gravity is 500 newtons and a lot of energy has to be transferred to lift the block upwards a distance of 1 metre.

Each time the same force of 500 newtons was acting, but the work done was small in one case and large in another.

To calculate the work we must make sure that we use the *distance moved in the direction in which the force acts.*

Questions for class discussion

1. If one joule of energy is transferred when raising a mass through 1 metre, how much energy is transferred raising it through 2 metres? Through 3 metres? Through 10 metres?

2a. If the strength of the gravitational field is 10 N/kg, what is the force exerted by gravity on a mass of 10 kg?

b. What is the work done, measured in joules, when the 10 kilogram mass is raised through 10 metres?

c. How much energy, measured in joules, has been transferred from chemical energy in your body to potential energy (uphill energy) of the mass?

d. How much energy would be transferred if the mass were 20 kg instead of 10 kg?

3. There are two shelves on a wall, one is 1 metre and the other 2 metres above the ground. Jack lifts a 1 kg mass from the ground on to the lower shelf, he rests a moment and then lifts it onto the second shelf. Jill lifts another 1 kg mass from the ground and puts it straight onto the top shelf. Did Jack do more work, the same work or less work than Jill?

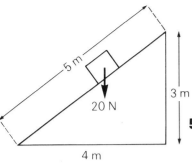

4. A block has a downward force on it due to gravity equal to 20 newtons. It slides down a *smooth* slope as shown on the left. The potential (uphill) energy at the top turns to kinetic (motion) energy near the bottom. In calculating the energy transferred, one boy says it is the force (20 N) times the distance (5 m), in other words, 100 J. Another boy says it is 20 N × 3 m, that is 60 J. Which statement is correct?

5a. A crate has a weight of 100 N. A man lifts it onto the back of the lorry which is 1 m above the ground. What is the potential (uphill) energy gained by the crate? Where does this energy come from?

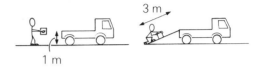

b. He finds it easier to get the crate on to the lorry if he slides it up an inclined plank of wood 3 m long. What is the potential energy gained by the crate this time? Where does this energy come from?

c. In fact the man will have to transfer rather more energy pushing the crate up the rough plank than lifting it straight onto the lorry. Where has this extra energy gone?

d. Why does he find it 'easier' to get the crate onto the lorry using the inclined plank?

Conservation of energy

You have seen many examples of energy being changed from one form to another. It is important to realise that you cannot create energy, nor can you destroy it. This is what scientists refer to as the *conservation of energy*.

We do not use up energy: when we do work we are transferring energy from one form to another. It is rather like money: when your father gives you some money, it is a transfer from his pocket to yours. Similarly when you buy something in a shop, there is a transfer of money from you to the shopkeeper, you have lost money and he has gained it. In the process you may have bought something in much the same way as work may be done in the process of transferring energy.

Human energy

How much energy do we need? This depends very much on our age and what we spend our time doing, but the following figures give some idea of what energy is needed per day. The energies are measured in kilojoules (kJ), and one kilojoule equals 1 000 joules.

Child (1 year old)	4 000 kJ
Child (5 years old)	6 000 kJ
Child (10 years old)	8 000 kJ
Boy (12 years old)	11 000 kJ
Girl (12 years old)	11 000 kJ
Boy (18 years old)	14 000 kJ
Girl (18 years old)	10 000 kJ
Adult man (light work)	11 000 kJ
Adult man (very heavy work)	20 000 kJ

The energy required for a man working ranges from 50 kJ to 10 kJ every minute depending on whether the work is heavy or light. Just resting in bed requires nearly 4 kJ per minute.

This energy comes from the food we eat and some idea of what is available from different foods is given below. In each case, the value given is the energy available from 1 kg.

Butter	32 000 kJ	Fried fish	9 000 kJ
Cheese	18 000 kJ	Eggs	7 000 kJ
Sugar	16 000 kJ	Potatoes	3 000 kJ
Beef	14 000 kJ	Oranges	1 500 kJ
Bread	10 000 kJ		

Questions for class discussion

1. How much energy would a man get from a plate of fried fish and chips? (This is an estimate question. First you must decide on the mass of the fish, then the mass of the potatoes. After that you can calculate the number of joules obtained from them.) ·

2. A man eats a breakfast of a glass of orange juice, two boiled eggs, two slices of bread and butter. How many joules does this give him?

   If all this energy is used in lifting sacks of potatoes a distance of 10 metres, how many sacks could he lift? Would he, in fact, be able to lift that number on such a breakfast?

Kinetic and potential energy

We first introduced kinetic energy by calling it motion energy, and potential energy by calling it uphill energy. Kinetic energy and potential energy are the more usual scientific names.

It may help to compare them with players and reserves in a game of football. The players are all tearing around the field, they have energy of movement and they represent kinetic energy. The reserves are waiting on the touch line, they are waiting to be given a chance to move, they are potential players and represent potential energy.

Kinetic energy can exist in various different forms. It can be the motion energy of a stone in flight. There is also rotational kinetic energy and this will be seen in Experiment 19.1. Furthermore, heat energy is really another form of kinetic energy. We have learnt already how solids are made up of atoms or molecules bound together. The atoms can vibrate and the hotter a body the greater the vibrations. So heat energy is really a form of internal kinetic energy of the vibrating atoms.

We are already familiar with uphill energy or gravitational potential energy. The strain energy stored in a piece of elastic when it is stretched is another form of potential energy. In Experiment 19.2 there is yet another form: in that case it is potential energy stored in a clock spring which has been wound up.

Since energy is so often being transformed from one form to another, it is wise to measure it in the same units, namely joules, whatever kind of energy it is.

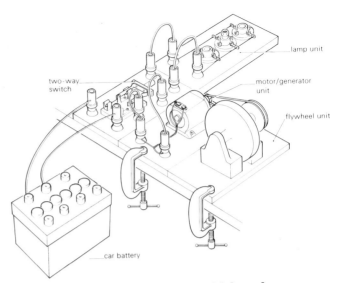

Various labels on the figure:
- lamp unit
- two-way switch
- motor/generator unit
- flywheel unit
- car battery

Experiment 19.1 Rotational kinetic energy

Set up the apparatus shown. Energy from the battery is transferred to the motor, which drives the flywheel. The flywheel has gained kinetic energy, but it is rotational kinetic energy, not quite the same as that of a ball thrown across the room, for the flywheel stays where it is!

If the switch is changed, the flywheel will drive the motor as a generator, which will light a lamp.

Experiment 19.2 Potential energy stored in a clock spring

Set up the apparatus shown. Energy from the battery drives the motor, which winds up the clock spring. The energy is stored as potential energy in the spring.

When the energy is released, it can be used to drive the motor as a generator and the lamp will flash momentarily.

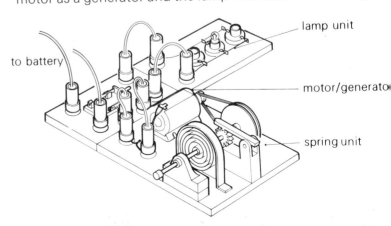

Labels on the figure:
- lamp unit
- to battery
- motor/generator
- spring unit

Machines and energy

To put a crate on the back of a lorry, it could be lifted on or pushed up an inclined plank. In both cases the same energy is transferred to the crate, but the plank enables us to do it more easily.

A machine is a device which enables us to do things more easily. A simple example of a machine is a lever.

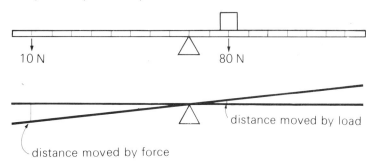

distance moved by load

distance moved by force

A mass of 8 kg is placed 2 units to the right of the fulcrum. The downward force on it due to gravity is 80 N. A force of 10 N, applied 8 units to the left, will balance this. A very slightly greater force will lift the load. Using the lever, it becomes possible to move the mass with a force which is slightly greater than 10 N. But 80 N would be necessary without the lever. It thus makes things easier. Is there any gain in energy?

Suppose the load is moved through 1 cm. How far will the 10 N force move? From the second drawing, you can see that it is eight times as far. The potential energy given to the load is

$$(80 \text{ N}) \times (\tfrac{1}{100} \text{ m}) \text{ or } \tfrac{8}{10} \text{ J}.$$

The work done moving the lever is

$$(10 \text{ N}) \times (\tfrac{8}{100} \text{ m}) \text{ or } \tfrac{8}{10} \text{ J},$$

the same result.

In other words, the lever makes it easier to move the load, but there is no gain in energy. The smaller force moved through the larger distance. A pulley system, see below, is yet another device to make things easier.

Experiment 19.3   Investigation of a pulley system

Attach a spring balance to a 1 kg mass. Because the gravitational field is 10 N/kg, the reading will be 10 N.

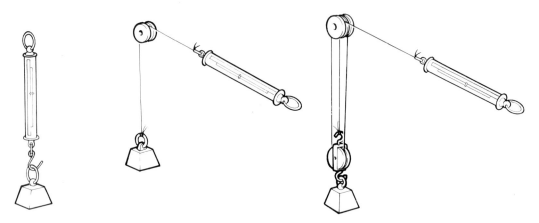

Set up a pulley as in the second figure above. The spring balance should still read 10 N. The effect of the pulley is merely to change the direction of the force.

Then set up the pulley as in the third figure. The reading on the spring balance is much less than 10 N, and by pulling on the string you can get the mass to rise with a smaller force than was necessary without the pulley system. Is there an increase in energy this time?

Find out how far you have to pull the string in order to raise the load 10 cm. Calculate how much energy is transferred to the load and how much energy is needed in the effort to pull on the string. Once again you should find that there is no gain in energy. A pulley system merely makes things easier.

Questions for homework or class discussion

1 a. Water from a reservoir runs down pipes to drive turbines. These turbines produce electricity which is used to drive a railway train.

b. At night power stations supply electricity to pumps. These pump the water back up to the higher reservoir.

List the energy changes involved at each stage, stating what form the energy is in.

Suggest the reason for using electricity to pump water to the reservoir when it might be used for heating our homes.

2. Give one example of each of the following energy transfers.
a. Kinetic energy to gravitational potential energy.
b. Potential energy to kinetic energy.
c. Kinetic energy to heat energy.
d. Chemical energy to heat energy.
e. Electrical energy to radiation (light) energy.
f. Potential energy to electrical energy.

3. A man has a mass of 70 kg. He walks up a hill 200 metres high.

a. If the strength of the gravitational field is 10 N/kg, what is his weight?

b. What is the gravitational potential energy gained in climbing the hill?

c. This energy comes from the chemical energy stored in the man's body through the food he has eaten. The actual chemical energy used in climbing the hill is about three times as great as the gravitational potential energy calculated in b. What has happened to the rest of the energy?

d. When he walks down the hill, what happens to the potential energy which was gained on the way up?

4. Use the details given on page 240 to work out how much butter, cheese, sugar, beef and eggs a heavy manual worker might need to eat in a day.

Chapter 20 **Heat**

12 volt battery

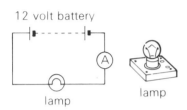

lamp

lamp

A car battery (12 volts) is connected to a car head-lamp bulb, which is marked 12 V, 24 W using the circuit on the left. The reading of the ammeter will be about 2 A.

If the lamp is replaced by one marked 12 V, 36 W, what do you notice about the brightness? It is certainly much brighter and you will notice that the ammeter now reads about 3 A.

In both these examples energy has been transformed from the battery to heat and light in the lamp, and the greater the current the more energy is transferred. Because heat is produced the lamp might be used to warm water if it were immersed in it. In fact this would be a dangerous thing to do as the water might short circuit the lamp. Later we will use special immersion heaters which are designed so that they can be put in water. See two immersion heaters illustrated below.

When heat energy is given to a body, it gets hotter. Doubtless you know from experience what that means. In fact our senses are not always very good at deciding whether something is hot or cold, as the experiment on page 61 showed. It is much more satisfactory to use a *thermometer*.

immersion heaters

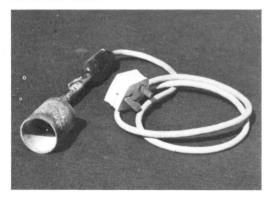

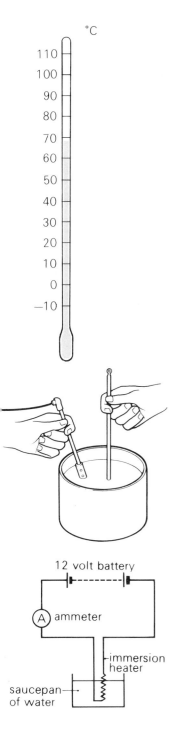

The commonest kind of thermometer which you will use in your laboratory is a mercury thermometer, marked in degrees Celsius (named after the Swedish scientist who suggested the scale). On this scale 0 °C is the temperature at which water freezes. 100 °C is the temperature at which water boils. Normal room temperature might be about 18 °C, a very hot summer day might be 30 °C.

Experiment 20.1 The heating effect of an immersion heater

a. Put a kilogram of water into an aluminium saucepan. Use a thermometer to measure the temperature of the water. Connect the 12-volt immersion heater to a 12-volt car battery and put it in the water. Find out how long it takes for the temperature of the water to rise 5 °C.

When doing this experiment, the water must be kept well stirred. Why is this necessary?

b. Pour away the water and do the experiment again, using only $\frac{1}{2}$ kg of water. How does the time compare with that needed for 1 kg?

c. How long do you think it would take to heat 1.5 kg of water through 5 °C? Decide what you think is likely to be the answer and then do the experiment and see if the result agrees with your forecast.

d. If you have time, it is an interesting experiment to put 1 kg of water in the saucepan, to switch on the immersion heater and to watch how the temperature rises. Keep the water well stirred and read the thermometer every minute. Plot the readings on a graph with the temperature along the vertical axis and the time along the horizontal axis. You will find the temperature rises steadily at first. Why does it not go on doing so?

Experiment 20.2 The energy transferred from the battery

Again connect the immersion heater to the 12-volt battery, but this time include an ammeter in series. The current is about 5 A. Use a watch to find how long the heater takes to heat 1 kg of water through 5 °C.

Now connect the heater to 6 volts instead of 12 volts. Notice what the current is, probably between 2.5 A and 3 A.

Again find how long it takes to heat 1 kg of water through 5 °C.

The current is just over half what it was originally, but it takes much more than twice as long to heat the water, in fact about four times as long.

Heating effect of an electric current

We know that the amount of energy transformed by a lamp depends on the current. The experiment with an immersion heater shows it depends on the voltage as well as the current. Later in your course, it will be shown that the energy transformed by a voltage, V, and current, I, in time, t, is given by

$$V \times I \times t.$$

For example, if the voltage is 12 volts and the current is 5 A, the energy transformed in 1 second is $12 \times 5 \times 1$ or 60 joules.

Experiment 20.3   Heating water

Put 1 kg of water into a saucepan. Connect the immersion heater to a 12-volt battery with an ammeter in series with it. Put the heater in the water and find how long it takes for the water to rise 5 °C.

Knowing the current, the voltage and the time, calculate the number of joules transferred by the immersion heater. This gives the amount of energy needed to raise 1 kg of water through 5 °C. From this, you can calculate the number of joules to raise 1 kg of water through 1 °C.

Your value is likely to be a little higher than the usually accepted value, partly because some of the energy may have been transferred to the surrounding air, partly because some will have raised the temperature of the saucepan as well as the water. More precise measurement shows that 4 200 J is needed to heat 1 kg of water through 1 °C.

Questions for class discussion

1. If 4 200 J is needed to heat 1 kg of water through 1 °C, how much energy is needed to heat

a. 2 kg of water through 1 °C?

b. 2 kg through 5 °C?

2. How many joules of energy must be transformed to heat 5 kg of water through 10 °C?

3. A kettle contains 1 kg of water at 0 °C. Estimate how many joules will be necessary to bring the water to 100 °C.

4. If a lamp has 12 volts connected across it and if the current flowing is 2 A, how much energy is transformed

a. in 1 second?

b. in 1 minute?

5. An electric fire is designed to operate from 200 volts and a current of 5 A flows through it. How much energy is transformed

a. in 1 second?

b. in 1 minute?

c. in 1 hour?

6. (Harder) An electric kettle operating from 200 volts takes 5 amperes. It is filled with 1 kg of water at 0 °C. Estimate how long it will take to bring the water to the boil (at 100 °C)?

Experiment 20.4   Heating aluminium

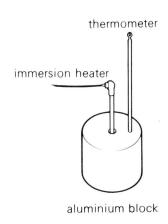

thermometer

immersion heater

aluminium block

You are provided with an aluminium block which has a mass of 1 kg. Insert a thermometer into it and measure the temperature. Again connect the immersion heater to a 12-volt battery in series with an ammeter. Put the immersion heater into the block and leave it there for a definite time, say 5 minutes. Remove the heater and measure the highest temperature reached by the aluminium block.

Calculate the energy transferred by the heater to the block. For example, if the voltage is 12 V, the current 5 A and the time 5 minutes (that is, 300 seconds), the energy transferred $= V \times I \times t = 12 \times 5 \times 300 = 18\,000$ J. Your value will be the amount of energy to raise 1 kg of aluminium through the temperature rise you measured. You can then calculate the amount required to raise 1 kg of aluminium through 1 °C.

This quantity (the energy to raise 1 kg of a substance through 1 °C) is given a special name in physics, the *specific heat capacity* of the substance concerned.

Some values for the specific heat capacity

Substance	Specific heat capacity (number of joules to heat 1 kg through 1 °C)
aluminium	877
copper	410
iron	486
lead	126
paraffin oil	2 134
mercury	140
water	4 200

Effects of heat

In Chapter 12 we discussed how heat could cause solids to melt, and liquids to turn to gases. We saw how this might be explained on a particle model of matter. The expansion of solids, liquids and gases is another important effect caused by heat. We shall look at this expansion in the following experiments.

Experiment 20.5   Expansion of solids

When a rod is heated, the expansion is very small and we must therefore use a special arrangement to detect it. The rod is fixed at one end using a 1 kg mass as shown. The other end of the rod rests on a needle. When the rod expands, the needle rotates and that makes the drinking straw turn. (It helps to put the needle on a piece of glass, perhaps a microscope slide, and a mass hung on the rod can stop slipping.) Heat the rod with a Bunsen burner and observe the expansion.

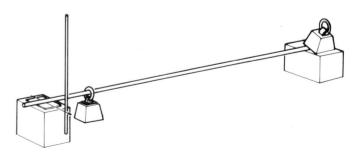

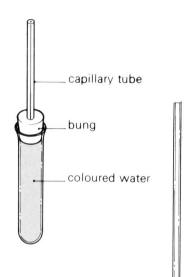

capillary tube

bung

coloured water

Experiment 20.6   Expansion of liquids

Fill the test tube with coloured water (a little ink is suitable for the colouring). Insert a bung with a capillary tube through it. Be careful to make certain there are no air bubbles left in the test tube.

Carefully hold the tube in a saucepan of nearly boiling water and watch what happens to the liquid.

Experiment 20.7   Expansion of gases

The flask contains air, trapped by the bung and the narrow bore tubing in which is a small bead of oil. Gentle heating of the flask with hands should produce sufficient temperature change to move the oil up the tube. What happens if the flask is placed in cold water, and then in warm water?

Experiment 20.8   Increase in the pressure of a gas when heated at constant volume

In order to keep the volume of the gas (air) constant, it is kept in a flask. The flask is connected to a Bourdon gauge. It is then put in a saucepan of cold water. See what pressure the gauge reads. Then heat up the water and watch how the pressure increases.

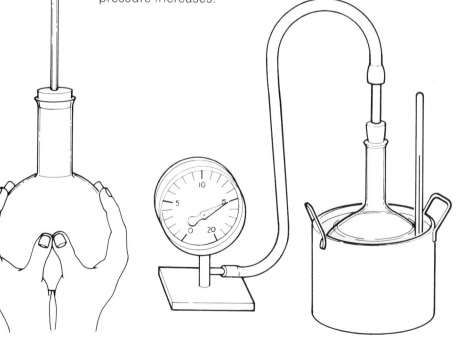

Homework assignments

There are many different books which will help you with these assignments. If you do not find what you want in the first book you try, then look in another. Looking in different books becomes a very necessary thing to do as you continue your studies in science. You will find that the topic book *Heat* by A. J. Parker and P. E. Heafford has much useful background material.

1. Prepare a short talk on how a mercury thermometer is made.

2. In this book we have talked about the Celsius temperature scale. Another scale is the Fahrenheit one. Find out how it differs from the Celsius scale.

3. Find out how a clinical thermometer (one used for medical purposes) differs from a laboratory thermometer.

4. When studying weather conditions, a maximum and minimum thermometer is often used. Find out how it works and prepare a short talk on it.

5. Those who build bridges or lay railway lines find that the expansion of solids can present problems. Find out how these are overcome and write a short essay on the topic.

6. Metals do not all expand by the same amount and a clever device is a bimetallic strip made of copper and iron fixed together.

   What happens to such a bimetallic strip when heated? Find out about ways in which use is made of this.

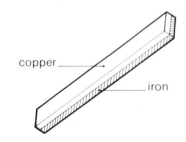

copper

iron

Heat flow

If you want to pass a written message from the back of a crowded hall to the front, you could pass it to someone in front of you, who then passes it to someone in front of him, and so on until the message reaches its destination. Another way of sending the message would be to give it to someone at the back of the room who pushes his way through the room until he delivers it in the front. Yet another way would be for you to throw the message from the back of the room

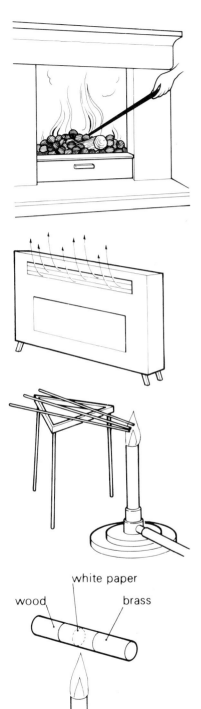

white paper

wood   brass

so that it reaches the front. There are three ways in which heat energy can be transferred, *conduction*, *convection* and *radiation*, and these have similarities with the three ways of sending the message.

If one end of a poker is put in a fire, the fire gives energy to the atoms in the poker so that they vibrate vigorously; in turn they will start their colder neighbours vibrating and so the energy is conveyed down the poker. This is *conduction*, in which heat in the form of motion energy is passed from neighbour to neighbour down the poker.

An electric convector heater in a room warms the air around it. This air expands and rises upwards because it is less dense, whilst more dense, cooler air comes in to take its place. This in turn gets warmed and rises. This stream of moving air is called a *convection current* and the process by which heat moves from one part of the room to another is called *convection*. The molecules move taking the energy with them.

Both conduction and convection require the presence of matter. As the space between the Sun and the Earth is virtually a perfect vacuum, heat cannot reach the Earth from the Sun by either conduction or convection. The process involved is usually referred to as *radiation*.

Experiment 20.9   Experiments on conduction

a. Support rods of various materials (copper, brass, aluminium, iron, glass) on a tripod as shown, so that one end of each is in the middle of a Bunsen flame. Wait until the rods are heated and then explore the temperature along each by sliding a finger carefully towards the hot end. In this way you can decide which rod conducts best. Alternatively you can move a match head along the rod instead of a finger and see when it ignites.

What happens if thinner rods are used? Does a thin rod conduct heat better or worse than a thick one?

b. Stick a piece of white paper around the join of a rod, one end of which is wood and the other brass. Hold the rod over a Bunsen flame as shown until the paper chars. Then look at the paper carefully: it appears to have charred where the paper was over the wood, but not over the brass. Why is this?

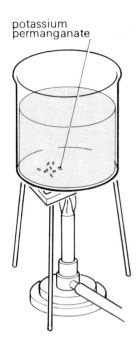

potassium permanganate

Experiment 20.10   Experiments on convection

a. Fill a beaker with cold water and put one or two small crystals of potassium permanganate on the bottom towards one side. Put the beaker on a tripod (without a gauze – the beaker will not crack if it is Pyrex) and heat it gently with a Bunsen burner. Convection currents will be seen in the liquid.

b. Fill a test tube with cold water. When the water is still put a single crystal of potassium permanganate at the bottom. Hold the top of the test tube in the fingers just above the water level and heat the bottom of the tube in a gentle Bunsen flame. Watch the convection currents and see how long you can hold the tube without discomfort.

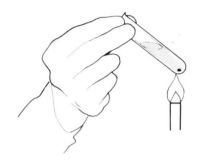 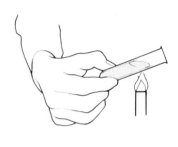

Empty the tube, wash it out and refill with cold water. Again put a single crystal of potassium permanganate at the bottom. This time hold the tube at the bottom and heat near the top of the tube, just below the water surface. This time there are no currents and you can hold the tube very much longer.

c. Convection in air can be shown with the apparatus shown. A lighted candle is placed under one of the chimneys. A piece of burning oily rag or a smouldering drinking straw is held at the top of the other chimney. The smoke will make the convection currents visible.

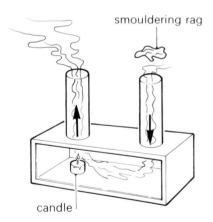

smouldering rag

candle

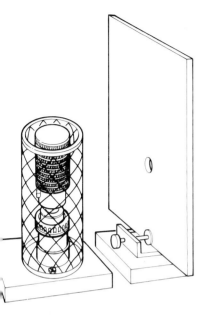

Experiment 20.11   Experiments on radiation

a. For these experiments the source of radiation is a heating element in front of which is an asbestos screen with a hole in it.

   Look at the element through the hole. Put the back of your hand in front of the hole to feel the radiation. Then put your cheek a short distance from the hole. Move your cheek further away and the heat energy received will be less.

   Insert a book between your cheek and the hole. The moment the book is put in the way the radiation is no longer felt on your cheek. Put a thin sheet of glass between the hole and your cheek, then two sheets to see what difference a thick sheet makes.

b. Heat a thick copper plate over Bunsen burners as shown. One side of the plate is dull black, the other shiny. Turn the plate sideways and hold your cheek near each side in turn. Much more heat appears to be radiated from the dull black surface.

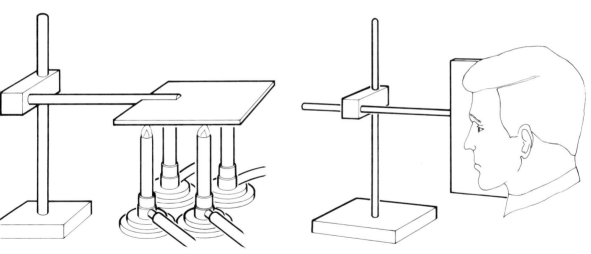

c. To investigate how different surfaces absorb radiation, first hold the back of your hand by the hole and feel the radiation for a short while. Then cover the back of the hand with aluminium leaf and then hold that by the hole. The hand can be held there much longer. Finally paint the aluminium leaf with *vegetable black*. When it is dry, put it in front of the hole once again. This time heat energy will be absorbed much more rapidly.

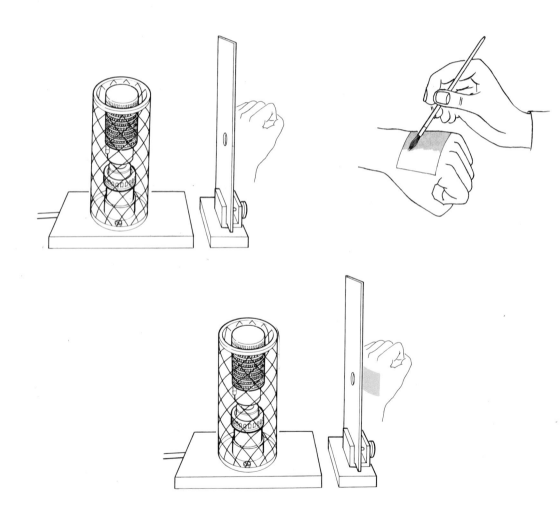

Questions for class discussion

1. A string vest is very good for keeping people warm, but it is full of holes. How do you explain this?

2. If you get out of bed with bare feet and stand on linoleum it feels cold, but standing on carpet makes you feel much warmer even though both the carpet and the linoleum are at the same temperature. What is the reason for this?

3a. If an electric immersion heater is being put into a hot water tank in the hot water system of a house, should it be put at the top or at the bottom of the tank?

   b. Why is the freezing unit inside a refrigerator put near the top of the refrigerator?

4. In the winter, a room usually feels warmer when the curtains are drawn. What is the reason for this?

5. The Earth is constantly receiving radiation energy from the Sun and it absorbs it. Will this mean that the Earth will go on getting hotter until it is at the same temperature as the Sun?

6. Why is a cloudy night usually much warmer than a clear cloudless one?

Homework assignments

1. Make a list of things in the kitchen at home which are good heat conductors, and also a list of those which are bad conductors.

2. Draw a diagram showing the hot water system of a house, including in it the heating unit, the storage tank for hot water, 1 bath and 3 basins. Add the cold water supply as well.

3. Write an essay on the various ways to improve the heat insulation of houses.

4. Sea breezes and land breezes at the coast are due to convection currents. Explain what causes them.

5. Find out how a vacuum flask (Thermos flask) is made. Draw a diagram to explain why it is made in the way it is.

Chapter 21

Revision tests E

Revision test E1

1. The size of an atom is about $\frac{1}{100\,000\,000}$ cm. If the atoms were put side by side in a line (like ball bearings touching each other), how many would there be in 1 cm?

2. Express the answer to question 1 in powers of 10.

3. Assuming the atoms are put side by side in a layer one atom thick, how many atoms would there be if the area of the layer is 1 cm²?

4. How many atoms would there be in a volume of 1 cm³?

5. A block of gold has a volume of 8 cm³. It is beaten into a block such that the thickness is everywhere 1 cm. What will be the area of this block?

6. Suppose the block in Question 5 is beaten into gold leaf $\frac{1}{1000}$ cm thick. What will now be the area?

7. If it were possible to beat the same block of gold into a layer one atom thick (in fact it is not) and if we assume the size of the atom is about $\frac{1}{100\,000\,000}$ cm, what will be the area?

8. Describe briefly how the oil film experiment enables you to make an estimate of the size of the olive oil molecule.

Revision test E2

1 a. A flat piece of metal P measures 20 cm × 20 cm. It has a mass of 10 kg and the downward force on it due to gravity is 100 newtons. It rests flat on the floor. What is the pressure on the floor? State what units your answer is in.

b. A flat piece of another metal Q measures 10 cm × 10 cm. It has a mass of 5 kg. What pressure does it exert when flat on the floor?

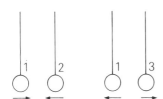

2. Three conducting balls are hung from insulating threads. It is found that when balls 1 and 2 are placed near each other they attract each other as shown. But when balls 1 and 3 are near each other, they repel one another. Give a possible explanation of this difference.

3a. Why is mercury, rather than water, more usually used in a barometer for measuring atmospheric pressure?

b. Draw a labelled diagram of a mercury barometer, shading in the mercury.

c. What do you think will be the approximate height of the mercury in the barometer tube?

d. Would you expect any difference between the pressure at the top and at the bottom of a high mountain? Give a reason for your answer.

4. A ball is released at the top of a slope. It has very little kinetic energy (motion energy) at the top of the slope, but by the time it has rolled to the bottom it has obtained a lot of kinetic energy even though no one has touched it. Where has this kinetic energy come from?

  The ball hits a rock at the bottom of the slope and comes to rest. Suggest what might have happened to the energy.

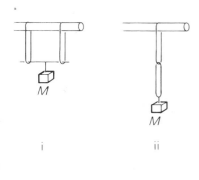

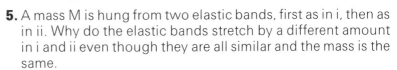

5. A mass M is hung from two elastic bands, first as in i, then as in ii. Why do the elastic bands stretch by a different amount in i and ii even though they are all similar and the mass is the same.

Revision test E3

1. Michael has a mass of 80 kg and he exerts a force of 800 newtons on the floor. The area of each foot in contact with the ground is 200 cm².

a. What pressure does he exert on the ground when standing on one foot? (Do not forget to give both a numerical value and the units.)

b. If he now stands evenly on two feet, what pressure does he exert?

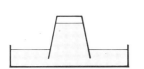

2. 'Water finds its own level.' A tumbler full of water with a glass plate over it is turned upside down. It is then put in a trough of water where the glass plate is removed. The levels stay as shown. Why does the water not find its own level?

3a. Draw a circuit diagram of a single cell lighting a lamp at normal brightness.
  b. Draw arrows on the circuit to show the direction of the current flowing in the circuit.
  c. Draw a circuit using a single cell and two lamps so that the two lamps glow with normal brightness.
  d. Mark on this circuit diagram, with a letter P, the point at which you would put a switch to turn out both lamps at once. Mark with the letter Q the point at which you would put a switch to turn out one lamp only.

4. A thin rubber balloon filled with hydrogen rises through the air. How do you explain this?
  Do you think that there is a limit to the height to which the balloon rises in air? Give a reason for your answer.

5. In the Brownian motion experiment smoke particles are illuminated and viewed through a microscope.
  a. Explain why smoke particles are used.
  b. When you look through the microscope why can you not see air molecules?
  c. What would happen to the Brownian motion if the cell containing the smoke were cooled?

Revision test E4
1. The cells and the lamps in the circuits are all similar.
  a. In which circuit will the lamp glow brightest?
  b. In which circuit will the cell or cells run down fastest?

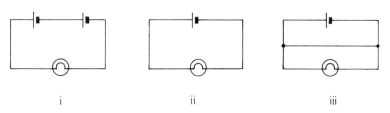

i   ii   iii

2. A boy hangs on a wire and it stretches 6 mm (Figure i). Using a second similar length of wire and a light piece of wood, he makes a swing.

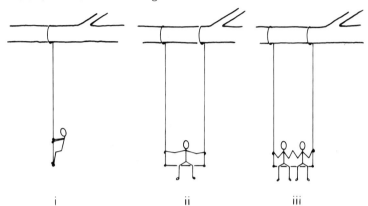

i   ii   iii

How much do you think each wire will stretch (Figure ii)? Give a reason for your answer.

A second boy with the same mass joins him on the swing (Figure iii). What will the stretch be now? Give a reason.

3a. A see-saw has marks on it equally spaced outwards from the central balance point. 10 kilograms are put on the left eight marks from the centre. 15 kilograms are put on the right four marks from the centre. Will these balance? If they do, explain why. If not, say which side goes down and why.

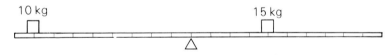

b. A beam is balanced at its centre point. A downward force of 20 newtons is then applied two units from the centre. What upward force *P* must be applied eight units from the centre if the beam is to remain balanced?

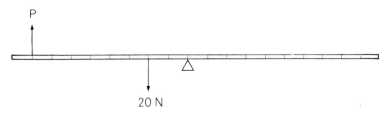

4. Explain both of the following.

a. A balloon is filled with air. When the balloon has been rubbed on a woollen sweater and placed in contact with the wall of a room, it stays there.

b. A boy walking across sand sinks in further if he walks on his heels than if he uses the soles of his shoes.

5a. What is meant by *density*? How would you show that the density of copper is about three times that of aluminium? You may assume that rectangular blocks of copper and aluminium are available in any size.

b. How would you show that a copper wire is a better conductor of electricity than an aluminium wire of the same length and thickness?

Revision test E5

1. This is a question about energy and energy transformation.

a. Scientists talk about *conservation of energy*. What does this mean?

b. When Moses went up Mount Sinai, he gained potential energy. Where did this energy come from?

c. On the top of the mountain he was given the tablets on which the Ten Commandments were written. Is it true to say that the potential energy of the tablets had all turned to kinetic energy when Moses stood with them at the bottom of Mount Sinai? Give a reason for your answer.

d. Describe what energy changes happened to Moses coming down the mountain.

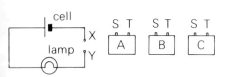

2. A, B and C are sealed boxes each having two terminals S and T which can be connected to X and Y in the simple circuit shown. The boxes are connected in turn, first with S to X and T to Y, then the other way round.

With A in the circuit, the lamp lights normally either way round. With B, the lamp lights equally dimly either way round. With C, the lamp goes out when the box is connected one way and shines very brightly when connected the other way round.

What might be inside box A? What might be in B? What might be in C?

3. Explain what is meant by conduction, convection and radiation. Give an example of each.

4. Suppose you have a collection of the following items : a plank of wood, some metal pipes and rods, some wires and pins, some rope and string, a 1 kg mass. You have no other masses and no weighing machines. Describe how you might attempt to find

a. your own mass,

b. the mass of a penny.

5. Several tin cans are *fastened* to the floor. There is a small hole in the top of each, from which comes a length of string. The lower ends of the strings are fastened inside the cans to various objects as follows :

a. a spring, the other end of which is fastened to the bottom of the can,

b. a 1 kilogram mass,

c. a magnet — and there is another magnet in the can fastened to the bottom,

d. a piston, almost as big as the inside of the can — and the can is filled with water,

e. a small iron ball — and the can is filled with sand,

f. the string is fastened directly to the bottom of the can.

　　You pull the strings upwards in order to find out what is in each can. Describe what you would feel on each occasion and how you might know what was inside.